CW00631259

1,000-500 million

Animals without backbones are multiplying in the seas. 800

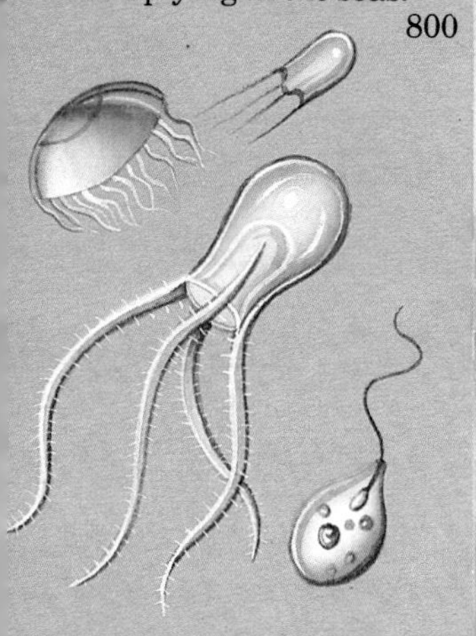

The oxygen level in the atmosphere reaches one-tenth its present level. 700

The ancient Atlantic Ocean is shrinking. 680

Ice covers central Africa. 650

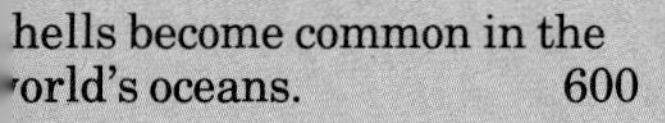

hells become common in the 'orld's oceans. 600

500-400 million

The first fishes and the first land plants and animals appear. 450-435

400-300 million

Extensive mountain-building and volcanic activity in Scotland, Wales and Scandinavia. 395-380

The first amphibians appear. 370

The first reptiles appear. 310

Chief coal-forming period, where partly-rotted vegetation accumulates in swamps. 310

300-200 million

Ice covers much of southern Africa, South America, Australia and Antarctica. 280

The ancient Atlantic Ocean closes. 250

Europe collides with Asia to form the Ural Mountains. 240

All continents join to form a super-continent, called Pangaea. 225

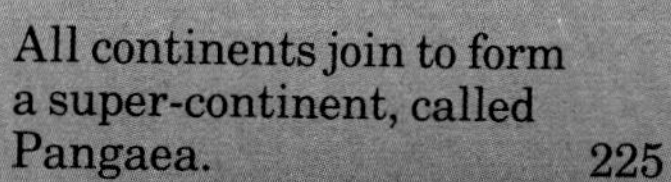

The dinosaurs appear. 210

Credits

Illustrations

Barbara Firth page 13, pages 14-15 (feature), pages 16-17 (feature), pages 18-19 (feature), page 27 (feature), pages 30-31 (feature), pages 34-35
Rory Kee end sheets
Ivan Lapper pages 38-39
Terry Pastor pages 2-3 (feature), pages 4-5 (top), pages 8-9 (feature), pages 20-21 (feature), pages 22-23 (feature), pages 32-33, cover
Tom Stimpson pages 6-7 (feature)
Tony White page 10 (left), page 11 (right), pages 36-37 (bottom)
Jane Wolsak pages 4-5 (bottom), page 7 (top), page 9 (bottom), page 12, pages 14-15 (top), pages 20-21 (bottom), page 23 (right), page 24, page 25, page 26, page 27 (bottom), pages 28-29 (bottom), page 29 (right), page 32 (bottom)
Elizabeth Wood page 3 (top), pages 10-11 (bottom), page 17 (top), page 18 (top), page 29 (top left), page 31 (top), page 37 (top), page 40

Editor Wendy Boase
Designer Pat Butterworth

Picture researcher Mary Fane
Photography
page 9 ZEFA/Starfoto; page 10 Institute of Geological Sciences; page 12 Institute of Goelogical Sciences; pages 24-25 Heather Angel; page 26 ZEFA/J.Bitsch; page 37 Pat Butterworth

Copyright © 1981 by Walker Books Limited
First published in Great Britain in 1981
by Methuen/Walker Books

All rights reserved. No part of this book may be reproduced or utilized in any form or by any means, electronic or mechanical, including photocopying, recording or by any information storage and retrieval system, without permission in writing from the Publisher. Inquiries should be addressed to Lothrop, Lee & Shepard Books, a division of William Morrow & Company Inc.,
105 Madison Avenue, New York, New York 10016.

Printed in Italy

First U.S. Edition

1 2 3 4 5 6 7 8 9 10

Library of Congress Catalog Card Number 81-80004

ISBN 0-688-00538-1 (pap.)
ISBN 0-688-00549-7 (lib. bdg.)

Contents

THE ACTIVE EARTH

Written by
David Lambert

Consultant K. Clayton
Professor of Environmental Sciences
University of East Anglia, Norwich

Lothrop, Lee & Shepard Books
New York

Earth takes shape

Imagine that a time machine could whisk you back more than 4,600 million years, to when the Earth was being formed. At first, you see only a huge spinning cloud of dust and gases. Gradually, the pulling force of gravity draws together the particles into a ball. This tugging crushes and heats the ingredients of the Earth until they melt.

Molten surface and poisonous air

To step outside your time machine, you need a fireproof suit with its own air supply, because this early Earth is like the inside of a hot, airless oven. Instead of sea, there is an ocean of molten rock. In some places, rocks have cooled and hardened into floating slabs. Some slabs have begun to stick together as islands of dry land.

But standing on these islands would be unpleasant. Streams of molten rock and clouds of poisonous gases and steam pour up from cracks and holes in the ground. Thunder deafens you, and lightning dazzles your eyes.

Millions of years later, the Earth would begin to look more familiar. As it cooled, the rock islands grew and joined, until the whole Earth had a crust of solid rock.

The familiar Earth

As the Earth's crust hardened, steam cooled to water vapour in the sky. Water vapour turned to rain which fell for many thousands of years. Rain-water filled huge hollows in the Earth's surface, so the first seas and oceans were born. Later, oxygen collected above the surface and helped to form the air that living things need to breathe.

Inside the Earth

All these changes happened only on the outside of the Earth. Scientists believe that the inside is still much as it was thousands of

millions of years ago. They think the Earth has four main layers. The thin outer layer is the hard crust of rock and soil that we live on. Deep down in mines, the crust is warm. Much deeper still, the rocks are often so hot that they can melt.

The light rocks of the crust probably float on a layer of hotter, heavier rocks, much as oil will float on water. This second layer is the mantle. Parts of it are soft enough to flow like very sticky tar.

Beneath the mantle lie the two layers of the Earth's core. The inner core seems to be a solid ball of metal. Around it is the outer core, a thick shell of liquid iron and nickel. The slow movement of these metals produces electric currents, which make the Earth behave like a giant magnet. Its opposite ends are the magnetic north and south poles.

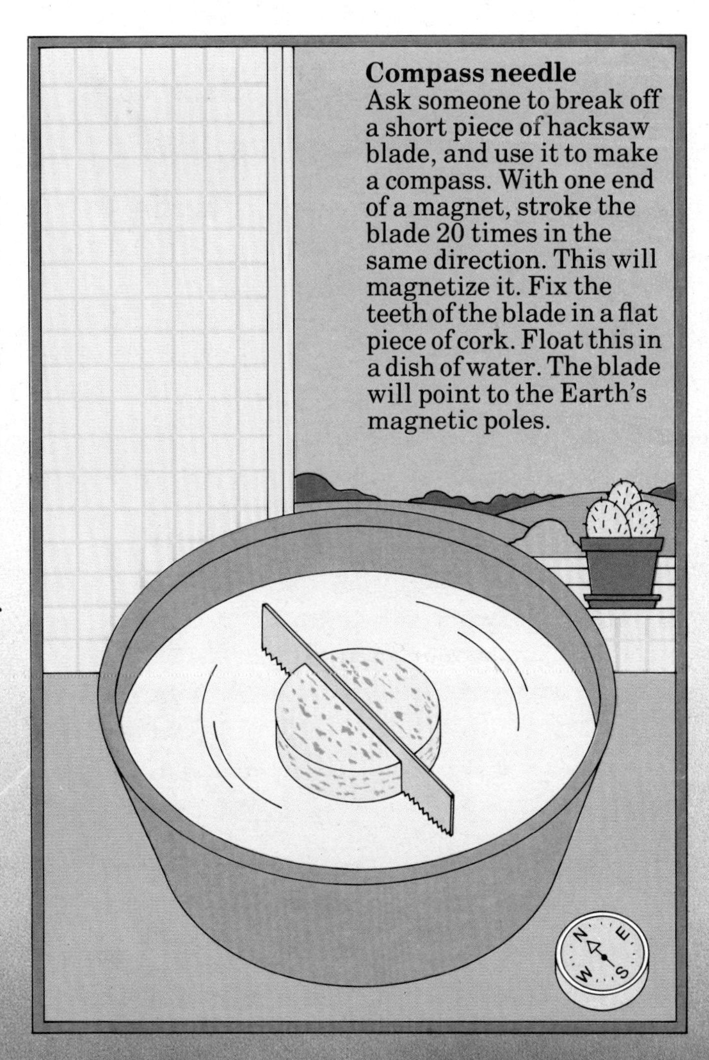

Compass needle
Ask someone to break off a short piece of hacksaw blade, and use it to make a compass. With one end of a magnet, stroke the blade 20 times in the same direction. This will magnetize it. Fix the teeth of the blade in a flat piece of cork. Float this in a dish of water. The blade will point to the Earth's magnetic poles.

The early Earth was a fiery mass of molten rock overhung by thick cloud, through which the Sun shone dimly.

Earth's four layers
The solid crust of the Earth formed at least 3,800 million years ago. It rests on a layer of heavier rock called the mantle. Beneath that is the liquid metal outer core. The inner core is a ball of solid, hot metal.

Lands adrift

When Christopher Columbus travelled from Europe to America in 1492, he sailed across thousands of kilometres of ocean. But if he had lived 200 million years ago, he could have made the journey in a single step, because at that time the continents of Europe and America were joined.

The continents

There are seven giant slabs of land called continents. Australia is the smallest. Next comes Europe. Antarctica is the fifth largest. The fourth largest is South America. North America ranks third. Africa is the second biggest. The largest continent of all is Asia, which is three times the size of Antarctica.

Pangaea – the super-continent

People used to think that the continents had always had the same sizes, shapes and positions. Then, in 1910, Alfred Wegener argued that all continents had once been joined as one great land mass. This German scientist named that super-continent 'Pangaea', which means 'All Earth'. Wegener believed Pangaea gradually broke into bits which drifted off to form the continents we know today.

Cracks such as this one in the East African rift valley may eventually split apart Africa, and form a new ocean.

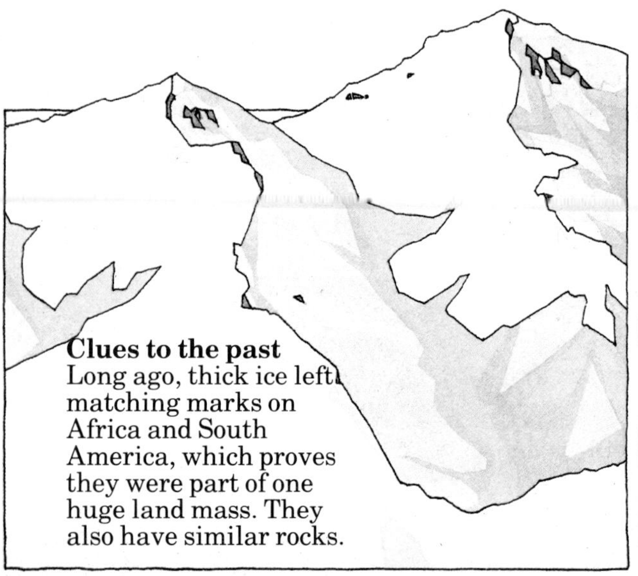

Clues to the past
Long ago, thick ice left matching marks on Africa and South America, which proves they were part of one huge land mass. They also have similar rocks.

Africa and Antarctica were also joined. The remains of an extinct hippo-like reptile, *Lystrosaurus,* as well as of the same seed fern, have been found on both continents.

This 'squashed' world shows that the Earth has at least 15 crustal plates. Cracks in the surface are the rims of plates that support the oceans and continents. Some continents look like pieces of a jigsaw puzzle.

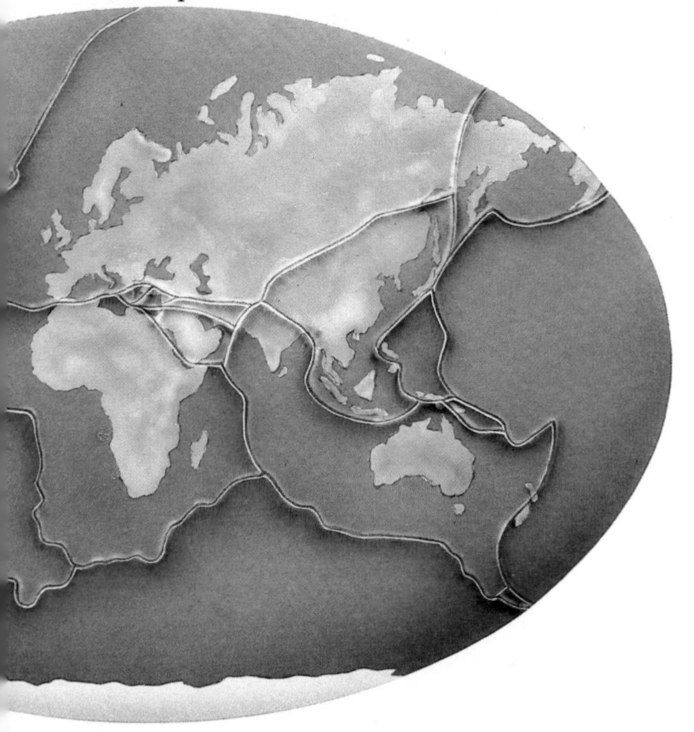

For many years, no one could understand how the continents could wander. Now it is believed that the Earth's crust is made up of at least 15 interlocking plates which are being slowly pushed and pulled around.

Pulling continents apart

Some plates carry oceans, while others bear continents. The 'engines' moving them are currents of sticky rock flowing through the mantle, beneath the crust. Where a rising current hits the crust, the current divides and spreads out, pulling the crust above in opposite directions. In this way, North America and Europe are slowly being pulled apart. Meanwhile, liquid rock wells up on the floor of the Atlantic Ocean, to fill the gap that keeps opening up between the plates that bear these two continents.

Losing ocean floor

In other places, the sea floor is being drawn down inside the mantle. In the eastern Pacific Ocean, a 15cm-strip of ocean floor thousands of kilometres long is sucked inside the mantle every year. This loss makes up for new sea floor being made in other parts of the Pacific and in the Atlantic Ocean.

Coal seams in Antarctica are the remains of trees that grew when this icy land lay in a warm area. It drifted to its present position after Pangaea began to break up.

Closely related animals and plants still thrive on lands now far apart. The ostrich of Africa, Australia's emu and the rhea of Argentina, for instance, may be related.

Earthquakes

The Portuguese city of Lisbon was one of the richest in the world, until it suffered an earthquake in 1755. As the ground began to shake, churches, palaces, shops and homes swayed like grass blown by the wind. Then they tumbled down. Soon, three waves as high as houses swept in from the sea. Fires broke out in the ruined city. As many as 60,000 people were burned or drowned, or crushed by falling buildings.

Each year, about half-a-million earthquakes shake parts of the world. Luckily, only one in 50 causes real damage. Most are so weak that people do not notice them.

Earthquake zones

Most earthquakes occur in clearly defined zones of the world. One zone runs round the Pacific Ocean, from New Zealand up to Japan, then down the west coast of the

Americas. In such places, two crustal plates meet each other. One may over-ride the other, or slide sideways past it, which sets up strains in rocks on land or under the sea. These strains may tear gashes, or faults, across the rocks. An earthquake is the result of the rocks on one side of a fault suddenly jerking down or up or sideways. Where the sea-bed shifts, giant waves called tsunamis may speed across the ocean.

A tsunami can travel faster than a train. Where it enters shallow bays, it rears up to drown villages and towns.

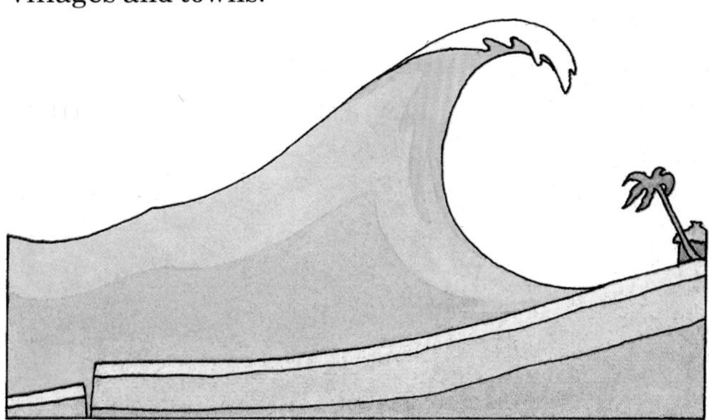

In different places, instruments called seismographs measure shocks arriving from one earthquake. By comparing these measurements, experts can find the epicentre – the area above the earthquake's focus, or starting point. The focus of some earthquakes is as much as 60km deep. But most have a focus of 20km or less, and among these are the ones that do damage on the surface.

Controlling earthquakes

Sometimes, earthquakes can be predicted. One clue is a gas called radon, which escapes from rocks that crack underground. The land may also bulge upward slightly. Since 1960 southern California has been bulging up near the San Andreas fault, a crack in the Earth's crust where two plates meet.

One idea for controlling earthquakes is to drill deep holes along a fault, and pump water into them so the rocks slide easily against each other instead of sticking. If this works, many small, harmless tremors will occur instead of a huge, destructive one.

Earthquakes are set off when rocks move beneath the crust. If the quake begins close to the surface, the shock waves will loosen trees and boulders, and cracks may open in the ground.

Volcanoes

Volcanoes are fiery mountains that grow out of the ground. Like earthquakes, volcanic eruptions are giant accidents that occur where the Earth's crust is weak. Most volcanoes grow where two of the 'jigsaw' pieces of the crust crash or separate. When this happens, underground molten rock (magma) and hot gases force their way to the surface through cracks or holes.

Erupting volcanoes

An erupting volcano is one of three kinds: explosive, quiet or 'in-between'. Explosive volcanoes behave with sudden violence. When the burning underground gases reach the surface, they rush out from the crater – the opening at the top of a volcano. Hot ash and cinders fly high into the air, and lumps of molten rock called lava drop like bombs upon the surrounding land. An explosive volcano usually builds a steep cone of ash.

With quiet volcanoes, lava flows gently from holes or cracks. Sticky lava hardens quickly and may pile up like a tower. If the lava is more liquid, it flows farther, building a gently sloping cone. Very liquid lava may flow over many kilometres. Old hardened lava from such flows has formed thick sheets over much of Iceland, Northern Ireland, India and western North America.

Quiet volcanoes may build gently sloping cones. In Hawaii, Mauna Loa's width is 50 times greater than its height.

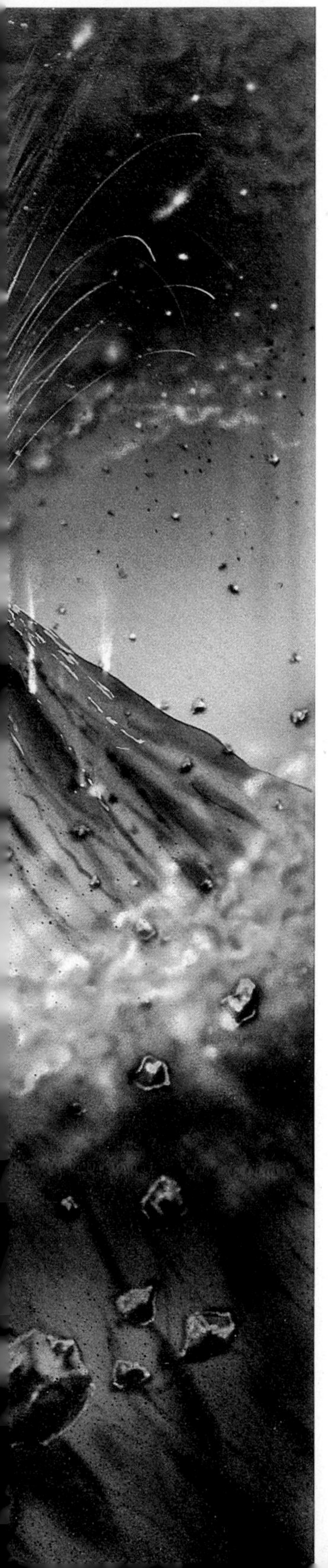

An explosive volcano hurls out hot gases and volcanic 'bombs'. Fine lava particles pile up as a steep cone of ash.

'In-between' volcanoes sometimes quietly pour out lava. At other times, they violently hurl out ash and gases. These volcanoes build cones in layers of ash and lava.

Extinct and sleeping volcanoes

Volcanoes do not erupt forever. Those never known to have erupted are said to be extinct. The Puy de Dôme in south-central France and the rock that Edinburgh Castle stands on, are both long-extinct volcanoes.

Not all inactive volcanoes are truly dead. Some are only dormant, or sleeping. Sometimes a solid mass of lava plugs a volcano's outlet. For hundreds of years hot gases collect below the plug, but nothing happens. Then, suddenly, the pressure of the gases blasts off the top of the volcano.

The hot Earth

Volcanoes are not only destructive. They build new land, and some volcanic rock slowly breaks down into fertile soil. Volcanoes also heat underground water, which surfaces as hot springs or as steam. Sometimes geysers of hot water jet above the surface. If the hot water is rich in mud, it builds a mud volcano. Some of these geothermal ('hot Earth') activities occur in the western USA, Iceland, Italy, Japan and New Zealand.

Where volcanoes are (or were) active, hot rocks heat underground water. If the hot water turns to steam deep down in a narrow crack, it forces water above it upward. The result is a geyser.

'In-between' volcanoes have cones that are built much like layer cakes. Layers of ash lie between layers of lava.

Fiery rocks

Somewhere or other, new rocks are always being added to the Earth's surface. Volcanoes, for instance, throw up material that forms new rocks. This helps make up for old rocks that are lost when pieces of the Earth's crust are pulled down inside the mantle.

Igneous rocks

Rocks formed from cooling magma (molten rock) are known as igneous ('fiery') rocks. All the rocks on Earth are igneous rocks, or ones made from particles in them. There are two main kinds of igneous rocks: one kind is formed underground, the other on the surface. The two are different, largely because of the rates at which they cool.

Slow cooling

When magma hardens underground, it cools slowly, and its ingredients form crystals large enough to see. Crystals are solid substances arranged in regular, geometric patterns. Various crystals can be seen in granite, a rock which forms underground. There may be grey, pink or white crystals of feldspar, glass-like quartz crystals and perhaps dark crystals of hornblende or shiny fragments of mica.

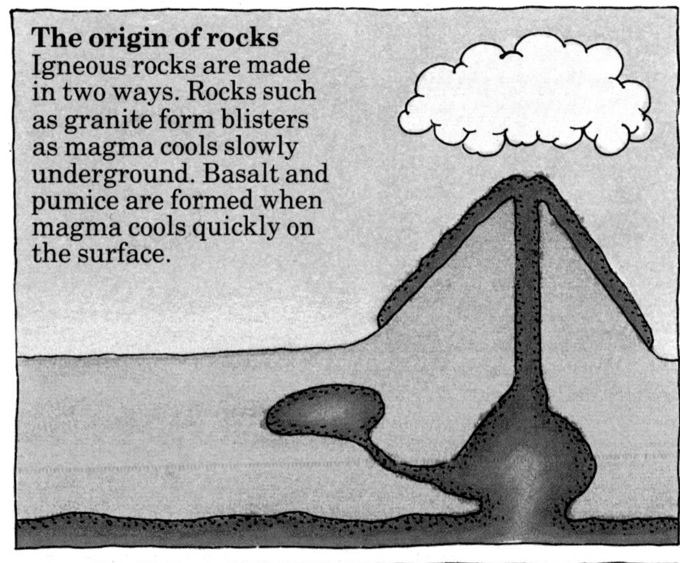

The origin of rocks
Igneous rocks are made in two ways. Rocks such as granite form blisters as magma cools slowly underground. Basalt and pumice are formed when magma cools quickly on the surface.

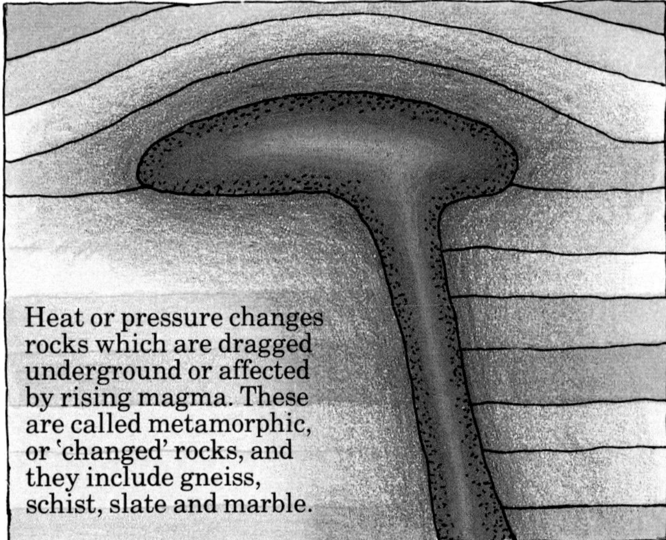

Heat or pressure changes rocks which are dragged underground or affected by rising magma. These are called metamorphic, or 'changed' rocks, and they include gneiss, schist, slate and marble.

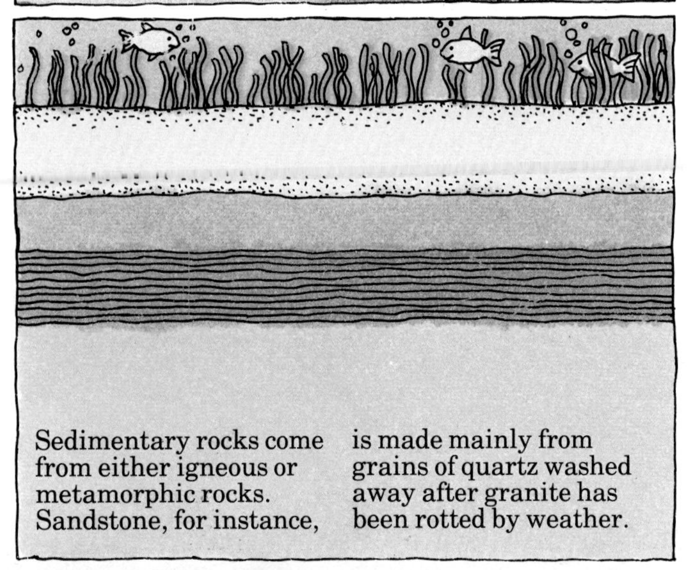

Sedimentary rocks come from either igneous or metamorphic rocks. Sandstone, for instance, is made mainly from grains of quartz washed away after granite has been rotted by weather.

Different minerals form differently shaped crystals. These milky quartz crystals have a pyramid form. The brown (iron-stained) dolomite crystals are cube-like.

As igneous rocks harden, they may fill cracks in older rocks or push them up into giant blisters. They show up on the surface only when rocks above them get worn down to a much lower level.

Fast cooling

When magma comes to the surface in volcanoes, it cools and hardens rapidly. It has time to form only small crystals. The volcanic rock basalt has tiny crystals. Obsidian, which comes from magma that cooled and hardened very fast indeed, has no crystals at all. It is smooth and shiny, like dark glass.

The Earth's rich store

Minerals are natural solids formed in the Earth's crust, and are made up of crystals. Diamonds, feldspar, quartz, silver and gold ores are just a few kinds of minerals.

Where people mine minerals depends on how and where these were formed. When magma cools as it bursts up towards the surface of the Earth, each mineral hardens into crystals at a different temperature. This is why miners often find copper, gold ores and silver ores in separate bands, sandwiched between various other kinds of minerals.

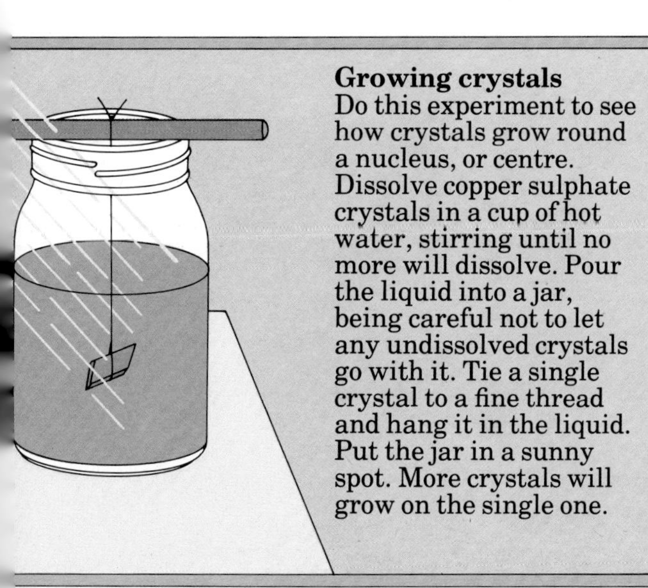

Growing crystals
Do this experiment to see how crystals grow round a nucleus, or centre. Dissolve copper sulphate crystals in a cup of hot water, stirring until no more will dissolve. Pour the liquid into a jar, being careful not to let any undissolved crystals go with it. Tie a single crystal to a fine thread and hang it in the liquid. Put the jar in a sunny spot. More crystals will grow on the single one.

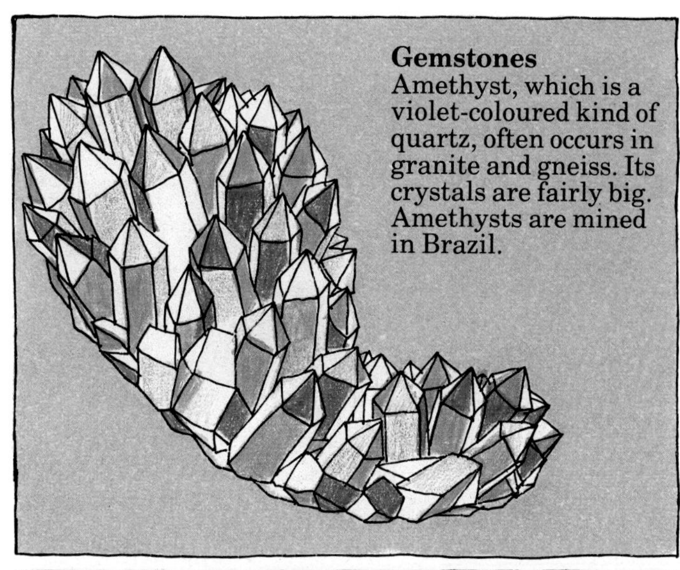

Gemstones
Amethyst, which is a violet-coloured kind of quartz, often occurs in granite and gneiss. Its crystals are fairly big. Amethysts are mined in Brazil.

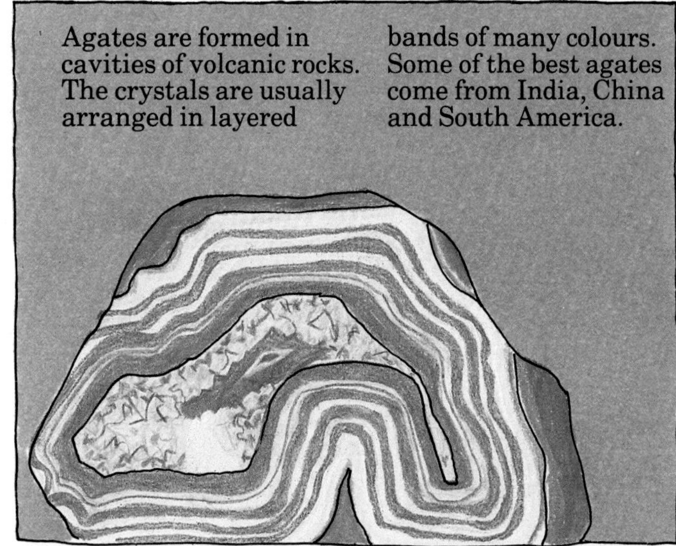

Agates are formed in cavities of volcanic rocks. The crystals are usually arranged in layered bands of many colours. Some of the best agates come from India, China and South America.

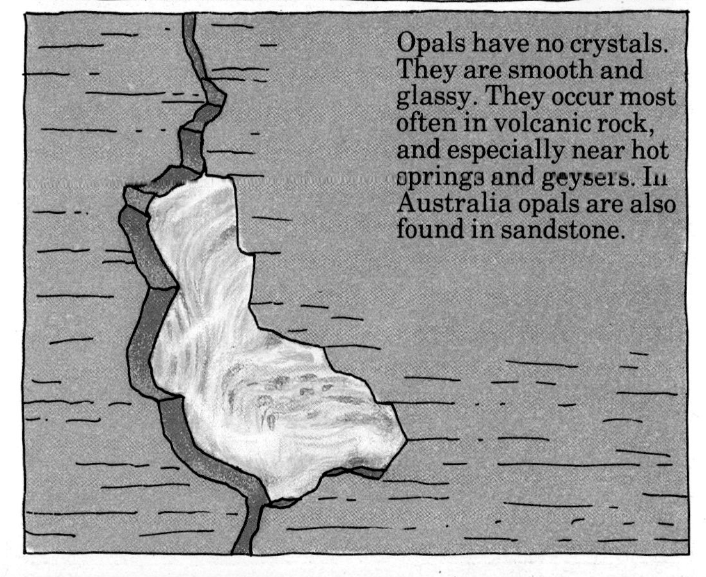

Opals have no crystals. They are smooth and glassy. They occur most often in volcanic rock, and especially near hot springs and geysers. In Australia opals are also found in sandstone.

Islands

In 1963, off south-west Iceland, the sea suddenly began to boil and churn. About a week later a brand new island rose up above the sea. People named this fiery heap of cinders Surtsey, after a fire god worshipped by the ancestors of today's Icelandic people. The new island was the tip of a volcano that had been growing upward from the bed of the sea, perhaps for centuries.

Volcanic islands

Like Surtsey, many islands are volcanoes that rose above the sea when hot rocks escaped from deep in the Earth's crust. This can happen along a crack in the crust where two plates are moving away from one another. In the South Atlantic, Ascension and Tristan da Cunha are islands of this kind.

Many other volcanic islands grow where one oceanic plate pushes into another and is run over. The front end of the lower plate is dragged down into the mantle and melts. But this melted rock from the Earth's crust is hotter and lighter than the mantle rock. So the melted crustal rock rises through the mantle, rather like a cork in water. Eventually, it bursts up through the floor of the ocean to build new volcanic islands. These islands often form curving rows called island arcs. The Aleutian Islands and many of the islands of Indonesia, Japan and the West Indies were formed in this way.

Drowned islands

In time, volcanic islands get worn down. Thousands of years ago, the level of the oceans rose and drowned many low volcanic islands. Others drowned because the sea-bed sank. Meanwhile, in warm, clear waters, coral grew upward round the volcanic islands as fast as they drowned. These coral islands are known as atolls.

Islands from continents

The world's largest islands are not volcanic in origin. They are chunks of land that once belonged to continents. Borneo, Sumatra and Sri Lanka belonged to Asia. New Guinea was once joined to Australia. The British Isles were part of Europe. All of them were cut off when great sheets of ice melted and raised the level of the oceans.

Some off-shore islands were never joined on to a mainland. Keys or cays are low islands built of coral sand or other particles piled up by the sea over many centuries.

Not all off-shore islands are large. Where waves chop through a cliff, they create stacks, or rocky islets, like these off the Shetland Islands.

The Hawaiian Islands rose one by one, as an oceanic plate moved over a fixed hot spot in the mantle. Every million years or so, molten rock pushed up through the plate and built a new volcanic island.

Where a volcanic island has drowned, a coral atoll may peep above the waves. Most of the coral was produced by coral polyps – creatures like tiny sea anemones. There are thousands of atolls in the South Pacific.

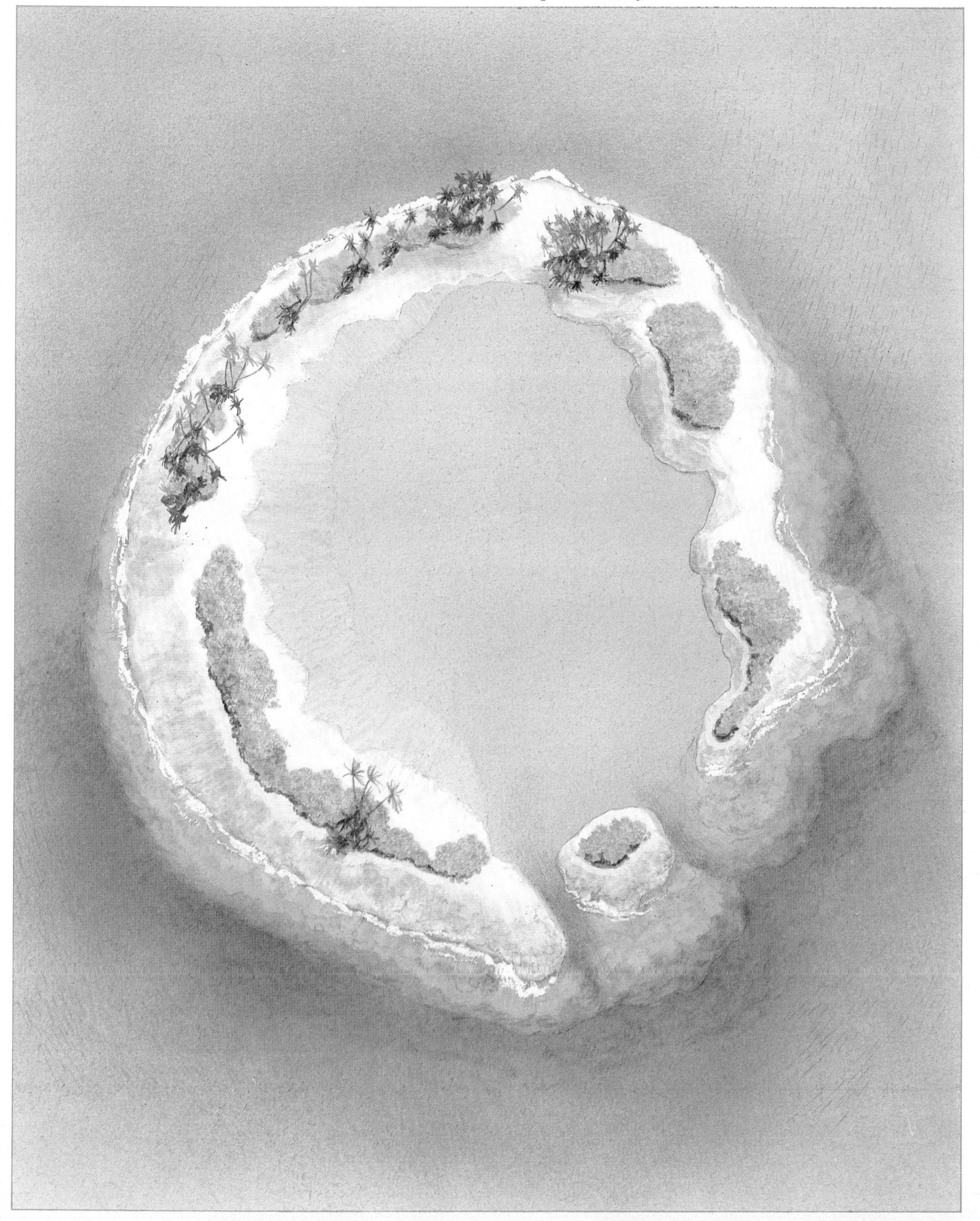

Mountains

Long ago, people climbing some mountains noticed the remains of sea shells embedded in the rocks. The climbers realised that these rocks high among the clouds must once have been beneath the sea.

The driving forces

The world's great mountain ranges are built of rock forced up several thousands of metres. Scientists believe this happened where two plates of the Earth's crust collided. The whole process was very slow. It can take between one and 10 million years to build a mighty chain of mountains.

Mountains folded by collision

The highest mountains on Earth, in the Himalayas, began with a grand collision between India and Asia no more than 50 million years or so ago. Before that, India was an island. Between India and Asia lay a shallow sea with a thick floor of layered rocks. When India slowly drifted north and pushed into Asia, these rocks were squeezed and crumpled. Flat layers of rock were forced up into giant humps, and into loops that toppled over. Mountains thrown up by this folding and buckling are known as fold mountains. The Himalayas are just a part of a great chain of fold mountains that runs from Spain to China. The chain includes the European Alps, which were pushed up when Italy glided slowly into southern Europe.

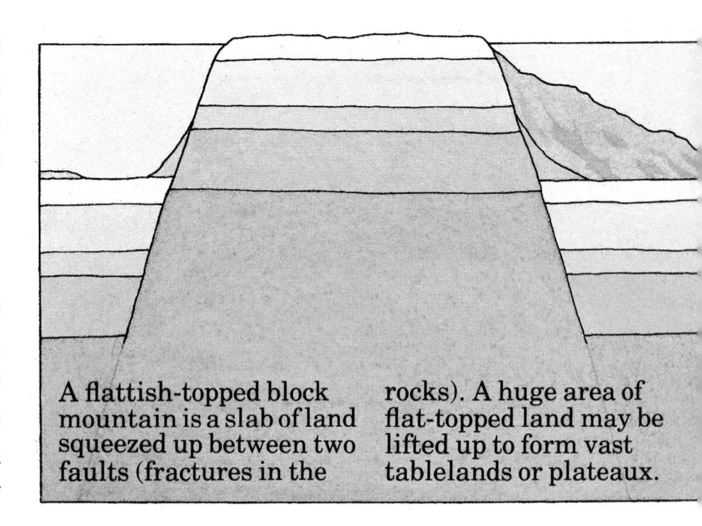
A flattish-topped block mountain is a slab of land squeezed up between two faults (fractures in the rocks). A huge area of flat-topped land may be lifted up to form vast tablelands or plateaux.

Another mighty mountain system runs down North and South America. It includes

If a slab of land slips down between two faults, it forms a rift valley. A great row of these runs from Syria, through East Africa, to Mozambique – almost one-eighth of the way round the world.

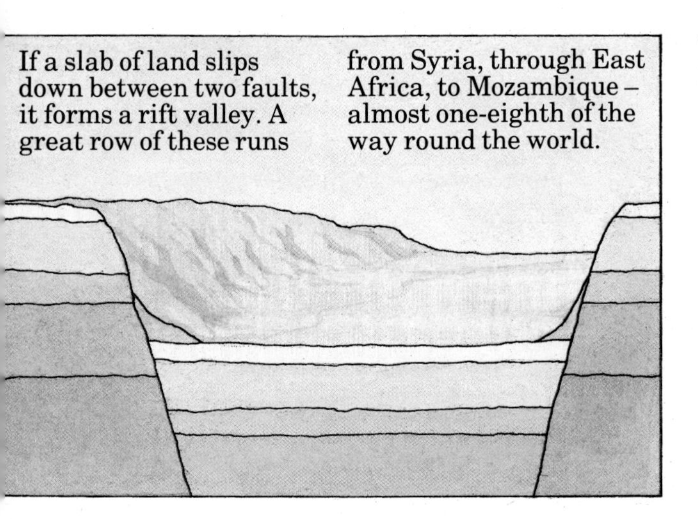

the Rocky Mountains and the Andes. The Andes is the world's longest mountain chain.

Volcanic mountains

Unlike the Himalayas, the Rockies and Andes formed when two crustal plates carrying continents crashed into two sea-floor plates under the Pacific Ocean. When these collisions occurred, much molten rock was forced to the surface and escaped to build volcanoes. Mt Aconcagua in the Andes is a volcanic mountain formed in this way.

Young and old mountains

Not all mountains are as young and tall as the Himalayas, the Rockies or the Andes. Some are so old that they have been worn down into low stubs. These include the Urals in the USSR, formed about 240 million years ago, and the Scottish Highlands, which are fold mountains about 400 million years old.

Fold mountains are made when plates collide, causing rocks to fold and buckle upward. Folding also heats and crushes the underground rocks, changing some of them into metamorphic ones.

Crumbling rocks

As soon as land is raised above the sea, powerful forces start to wear it down. In most places, the land is being attacked too slowly for changes to be noticed. But there are many clues to show what is happening.

Weaknesses in rocks

The attack upon the land begins when weather starts to break up solid rock. All kinds of rocks have weaknesses. Some, for instance, are soft. Some have millions of tiny gaps between their particles. Others are made of layers or blocks that can be prised apart slowly, and certain rocks contain particles that will dissolve in water.

The attack by frost

In cool, moist climates, rock is broken up by water that keeps freezing and thawing. When rain falls or dew forms, water fills cracks in the rock. If it freezes, frost and ice press hard against the sides of each crack. On high mountains, water may freeze each night and the ice may melt each day. This process can happen many times a year. So the ice gradually widens the cracks in the rock. In time, big boulders split in two, and the outer layers of layered rocks flake off. Stones broken from the mountain tops may fall and settle on the steep slopes below. Loose layers of these stones are called scree.

Battering from the Sun

In hot, dry places such as the Sahara Desert, hot days and cool nights help to start rocks breaking up. By day, heat makes the rocks try to expand. At night, they cool and try to contract. Expansion and contraction widens cracks in the solid rock until it breaks up into smaller blocks. Sometimes, pieces of rock

break off after sunset with a loud bang like a pistol shot.

Chemicals that dissolve rock

Chemicals in rain-water simply rot some rocks away. Limestone, for instance, is dissolved by rain and soil containing carbon dioxide. Water also rots feldspar, an ingredient of granite. Hard feldspar crystals rot into kaolin, a soft clay powder which potters use for making porcelain. The other minerals in granite are left as loose grains of sand. Some boulders rot until they are hollow shells. In Sardinia, farmers sometimes use the shells as stables and storerooms.

Besides weather, plants also help break up rocks. Roots widen cracks in them, and some lichens produce acids that dissolve particles in rocks. The top 1m or more is so changed that it is called soil.

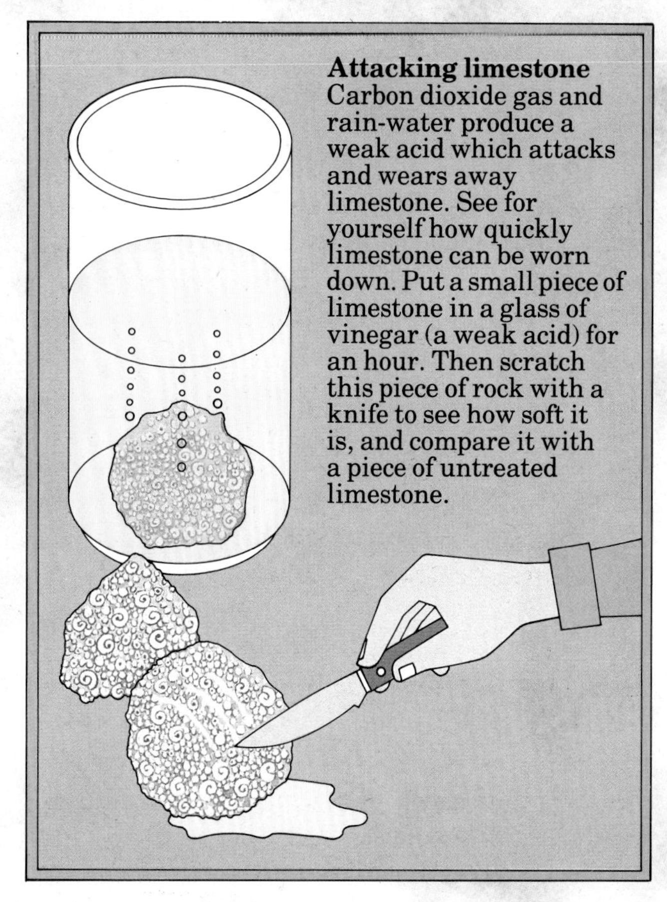

Attacking limestone
Carbon dioxide gas and rain-water produce a weak acid which attacks and wears away limestone. See for yourself how quickly limestone can be worn down. Put a small piece of limestone in a glass of vinegar (a weak acid) for an hour. Then scratch this piece of rock with a knife to see how soft it is, and compare it with a piece of untreated limestone.

Rain can rot limestone until it looks like rows of pavements separated by deep, narrow gutters. Such limestone hills occur in England's Peak District, in the Karst of Yugoslavia and in the Causses of France.

Rivers

Rivers are nature's chief removal men. They carry broken bits of weathered rock downhill towards the sea. They also carve out valleys. In time, they help wear away hills and mountains. Without the action of rivers, the land would look quite different.

Birth of a mountain stream

Many rivers start where spring water bubbles up from underground, then runs downhill as a stream. If the slope is steep, the stream flows fast. Stones, broken from the solid rock by weathering, are carried with it. These stones grind against each other. Grinding breaks them into particles of gravel, sand and mud. The stones also grind against the rocky stream bed and deepen it. Gradually, a young mountain stream cuts out a deep, steep-sided valley.

Waterfalls and rapids

The stream wears away soft rocks such as clay. But harder rocks, such as granite, are not so easily worn down. Where both kinds lie near each other, hard rocks are left as steps, and the stream pours over them in waterfalls and rapids.

The river at work

A young stream's cutting power gets greater as other streams, or tributaries, flow into it. Their added water swells the stream into a river. Eventually, after thousands of years, the river and its tributaries wear down mountains into a maze of valleys and hills.

Breaking down rock
Moving water helps break up solid rock. You can prove this if you test pieces of soft rock such as shale or sandstone. Wrap the rock in a cloth. Hammer it until it is broken into many pieces, each about the size of a fingernail. Put aside two pieces. Put the rest into a bottle half-full of water. Shake the bottle 100 times. Take out two or three pieces. Continue shaking and removing bits of rock. Compare pieces shaken many times with those taken out at an early stage. Then compare the shaken pieces with the two not shaken at all.

The bed slopes less steeply than before and, every year or so, the river floods the land on each side of it. A river such as this is said to be middle-aged.

Unlike a lively young mountain stream that can move big stones and even boulders, a middle-aged river shifts mostly sands, gravels and smaller particles of ground-up rock.

Where the river flows into a sea without strong tides, it may split up. The mud it drops builds a delta out into the sea.

Young streams carry big stones, and cut deep, steep-sided valleys. Ridges are left between the valleys.

It no longer digs down into its bed so quickly. Instead, the river wanders to and fro across the flattened valley floor. Its wide, 's'-shaped bends are called meanders.

The river in old age

A river reaches old age when it has completely worn down hills and mountains. The gentle slopes now yield few particles larger than sands and clays, and these are the river's main load. Where the river meets the sea, it drops this load. In sheltered estuaries, the mud builds up tidal marshes.

Some rivers split up to form deltas as they approach the sea. Deltas are places where rivers build new land out into the sea with the remains of rocks from hills and mountains that they helped to destroy and carry away.

Gradually, bends widen and flatten the valley floor. An old river meanders over a plain that is almost flat.

Middle-aged rivers form flat valleys by dumping sediment on the inside of bends, while gnawing away the outside banks.

Caves

Valleys are proof that rivers are wearing down the land. But some rivers attack rocks invisibly, below the ground. These underground rivers can hollow out huge caves.

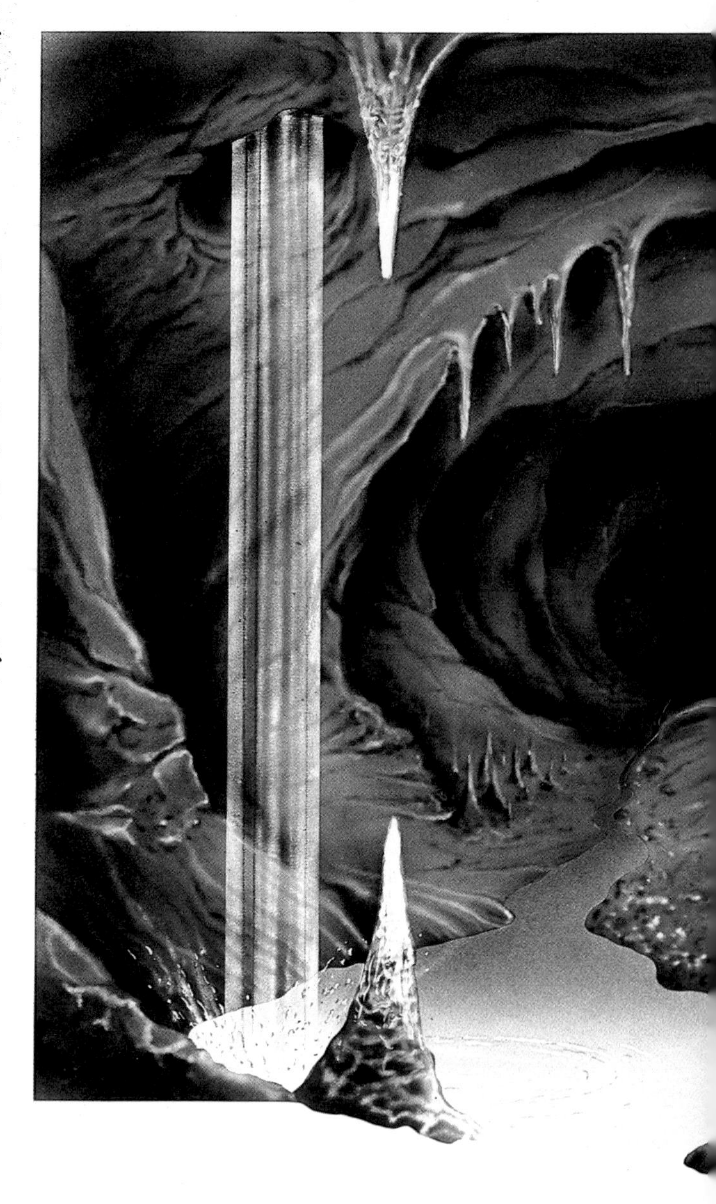

Dissolving mountains

The insides of many limestone mountains are easily dissolved to form caves. This is because their rocks are like a stack of giant building blocks, with narrow cracks between them. Some cracks run across the rocks. Others lead straight down, and where a surface stream finds such a crack, the water widens it. This grows into a hole big enough to swallow up the stream.

In some cases, many streams burrow into a mountain. A slice cut through the mountain would show it to be as full of caves as a slice of Gruyère cheese is full of holes.

Exploring caves

Thousands of years ago, underground rivers carried more water than they do today. Some have dried up altogether, which means that people can explore the caves that the streams created. Caving is cold, dark, damp and dangerous. Cavers use lamps, rope-ladders, and sometimes diving suits. They often climb, swim or wriggle through narrow passages.

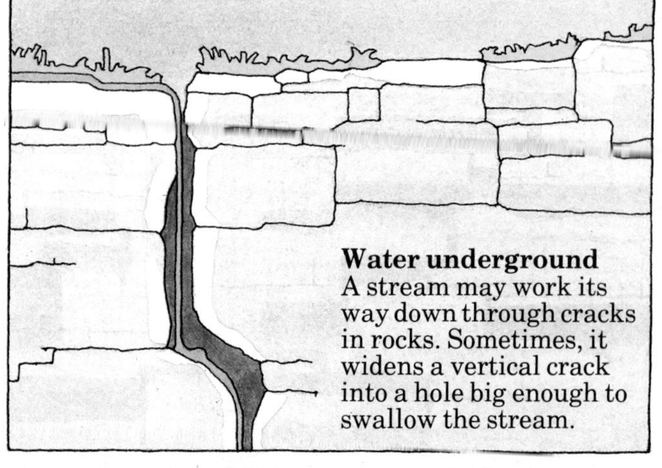

Water underground
A stream may work its way down through cracks in rocks. Sometimes, it widens a vertical crack into a hole big enough to swallow the stream.

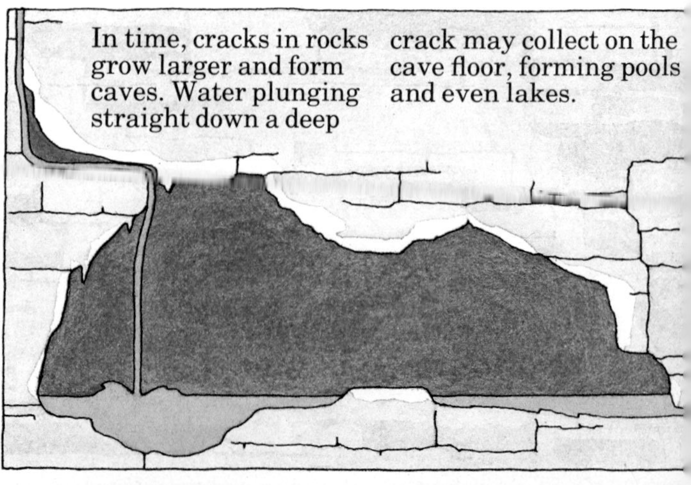

In time, cracks in rocks grow larger and form caves. Water plunging straight down a deep crack may collect on the cave floor, forming pools and even lakes.

Limestone caves are cut and shaped by the action of water. Stalagmites, built up from the floor, are usually shorter and thicker than stalactites, which hang from the roof. Where stalagmites and stalactites meet, they form a solid pillar.

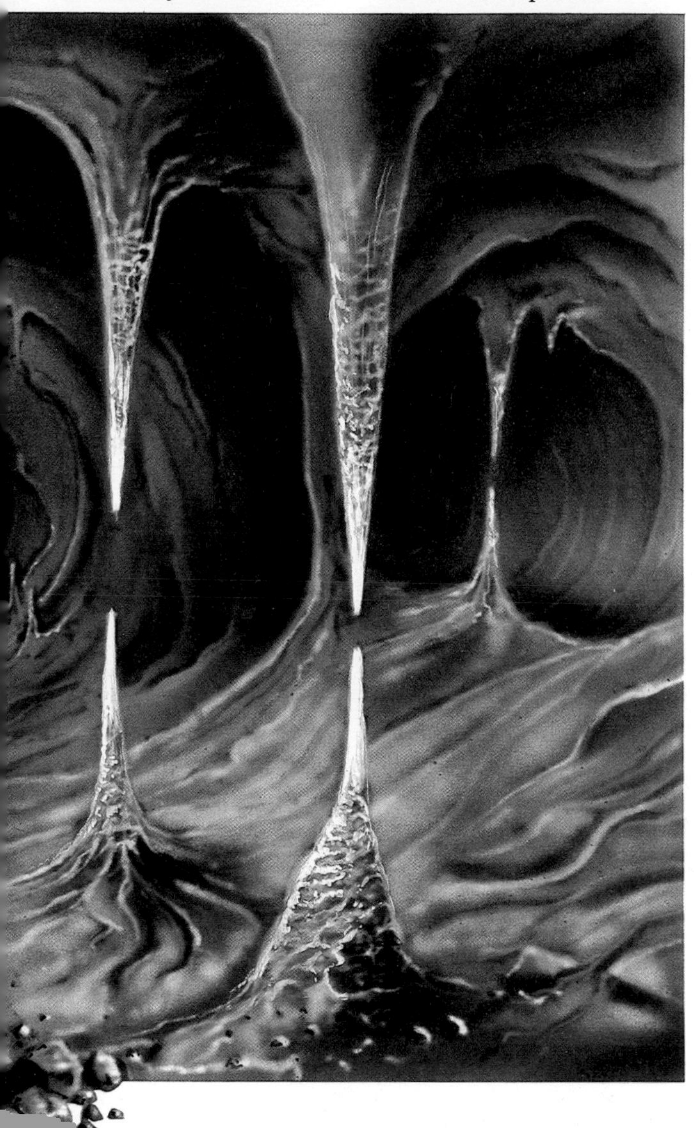

Cavers have made some astonishing discoveries. In France, they discovered two caves about 1,300 m deep. In Kentucky's Mammoth Cave National Park, American cavers found caves stretching over a distance of more than 300 km. This is the world's largest cave system. The USA also has the largest cave on Earth, the Big Room in New Mexico's Carlsbad Caverns. It is as large as five full-sized soccer pitches.

Stalactites and stalagmites

Many caves resemble the inside of some fairy palace. Shining spikes called stalactites hang down from the roof like icicles. Stalagmites are spikes which jut up from the floor. Both are made from calcite, the main mineral that water dissolves from limestone. When the water drips from the cave roof, or splashes on the floor, some of the water evaporates, or turns into water vapour. Evaporating water leaves a thin layer of calcite clinging to the rock. Eventually, calcite layers build a stalactite or stalagmite.

Stalactites grow slowly – about 1cm in 100 years. The longest calcite column known in the world is in a cave in southern Spain. This column is nearly 60m tall, and must have taken thousands of years to form.

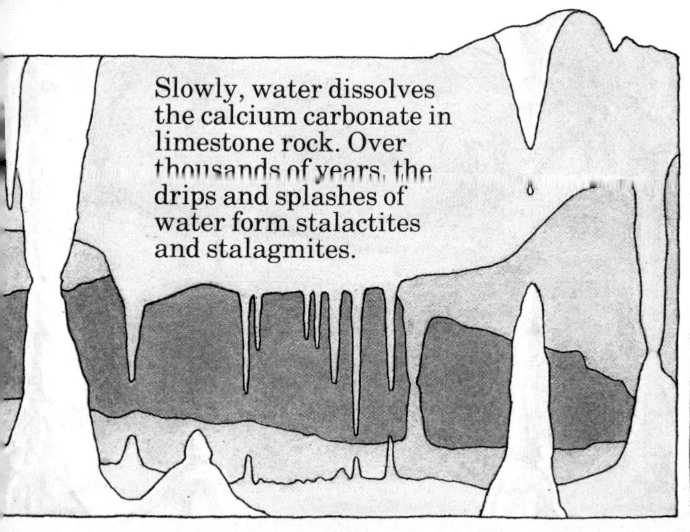

Slowly, water dissolves the calcium carbonate in limestone rock. Over thousands of years, the drips and splashes of water form stalactites and stalagmites.

A stream may zig-zag far underground, then flow out at the foot of a mountain as a spring. Dye put in water where it disappears enables its journey to be traced.

Invading ice

Millions of people now live in Norway and Sweden. But these Scandinavian lands were once dead and empty, for they lay deeply buried under ice. The great Scandinavian ice sheet melted more than 10,000 years ago, but before it disappeared it greatly changed the land it covered. Austria and Switzerland are two other countries where ice shaped the hills and valleys.

While Europe froze, another mighty sheet of ice crept out over northern North America. Today, a great ice sheet still sits on the southern continent of Antarctica. Another covers most of Greenland.

Glaciers

In these lands, and on many mountains in the world, ice rivers known as glaciers have gnawed away the rocks. Glaciers are born on mountain slopes. Firstly, snow collects in a hollow. In a big, deep snow hollow, snow at the bottom gets compressed and turns to ice. Pressure eventually forces out a tongue of ice, which starts to creep downhill. This tongue is a baby glacier.

On the far side of the mountain another hollow may also be growing larger. When two hollows meet, only a sharp mountain ridge is left between them. If three hollows meet, they may gnaw the mountain into a pyramid.

The glacier flows down a valley already cut out by a river. It grows larger as other glaciers flow in from side valleys. Norway's Jostedalsbre Glacier is 100 km long and one Antarctic glacier stretches over 500 km.

A glacier shapes its valley with stones and rocks stuck in the sides and bottom. They scrape against the valley sides and floor, rubbing away solid rock. The moving ice chops off ridges that jut into the valley, eventually widening and deepening it.

Glacial rubble

In cold countries glaciers flow down to the sea, where chunks break off and float away as icebergs. In warmer lands, the fronts of glaciers melt when they reach warm air, perhaps half-way down a mountainside. Stones that had been frozen in a glacier drop from its melting snout like lumps of coal reaching the end of a conveyor belt. The pile of rubble left behind is called a moraine.

Spreading the moraine

If the climate grows warmer, the snout retreats, leaving heaps of stones and sand behind. Streams flowing from the snout may spread this rubble. Sweden and much of northern Germany is covered by layers of clay, sand, gravel, stones and even boulders, which were brought by the great Scandinavian ice sheet and left behind as it melted.

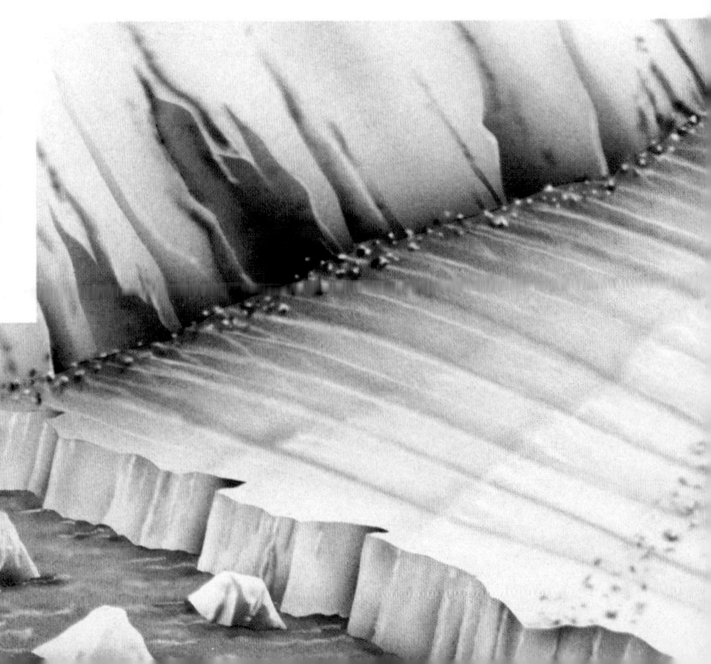

One pace a day is faster than most glaciers move. The sides often travel more slowly than the middle. This difference sets up strains, which causes deep cracks, or crevasses, to split the surface of the ice.

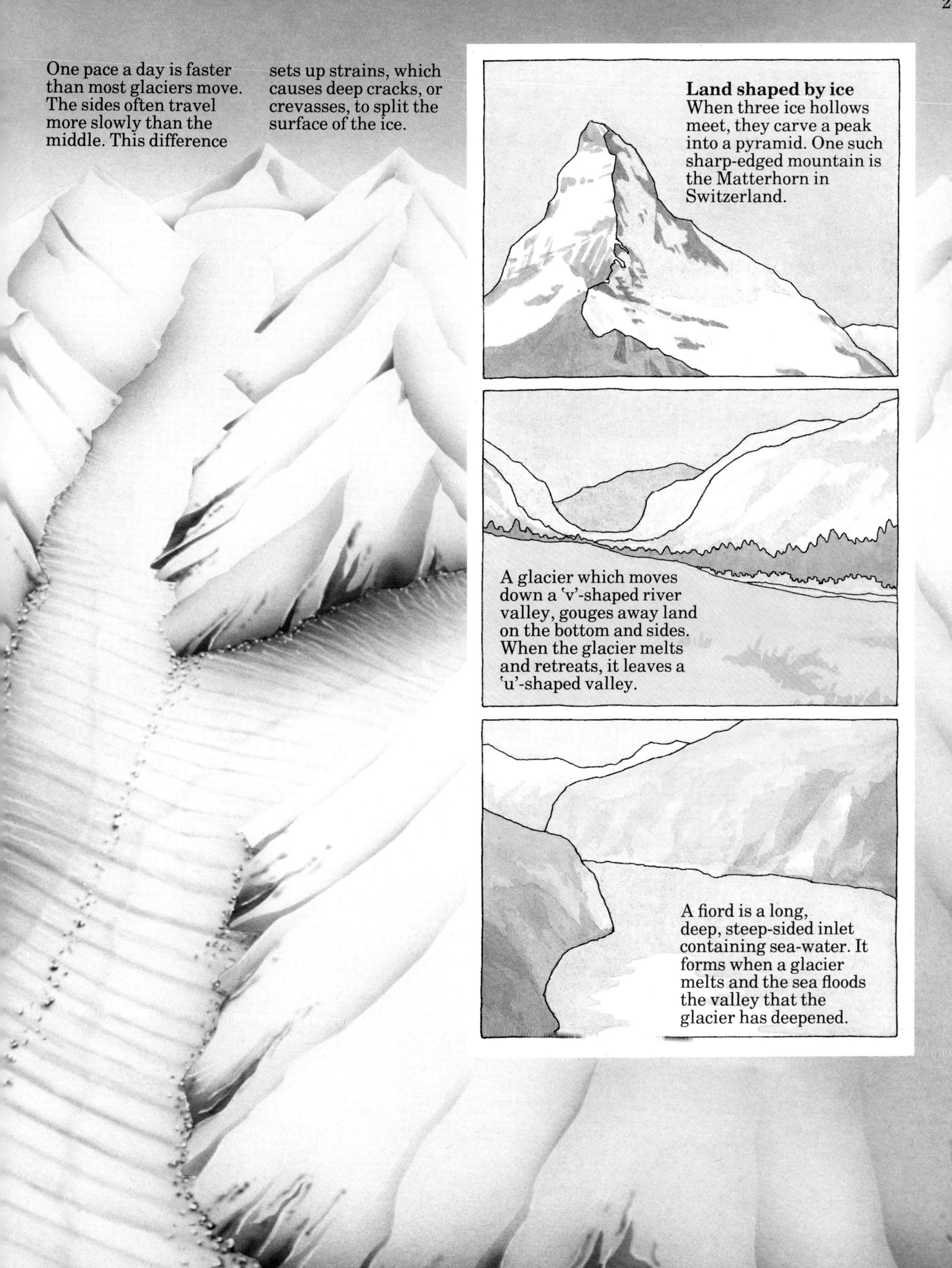

Land shaped by ice
When three ice hollows meet, they carve a peak into a pyramid. One such sharp-edged mountain is the Matterhorn in Switzerland.

A glacier which moves down a 'v'-shaped river valley, gouges away land on the bottom and sides. When the glacier melts and retreats, it leaves a 'u'-shaped valley.

A fiord is a long, deep, steep-sided inlet containing sea-water. It forms when a glacier melts and the sea floods the valley that the glacier has deepened.

Lakes

Lakes form where water flows into hollows or dips in the ground. Some lie in rock basins scraped out long ago by moving ice. Finland has 60,000 such lakes, covering 11 out of every 100 sq km of the country.

Most Finnish lakes are small, but some of the world's lakes are as big as seas. The largest of all lakes is the Caspian Sea in the USSR and Iran. This salt lake is almost as big as Japan. The largest body of fresh water is Lake Superior in Canada and the USA. It covers an area greater than Belgium.

Finger lakes

Some lakes fill valleys deepened by glaciers and blocked at the lower end by a wall of stones, sand and mud, called an end moraine. End moraines are built of rubble left by a melting glacier. When the entire glacier has melted, the water left behind piles up behind the end moraine and forms a long, narrow lake shaped rather like a finger.

Vast depths of water

Some of the largest and deepest lakes fill hollows left where the surface of the Earth has sunk. Among large lakes filling parts of the East African rift valley system are the Dead Sea in the Middle East, and Lakes Tanganyika and Malawi in East Africa.

Finger lakes may form when mountain glaciers melt. Many finger lakes lie in Swedish valleys and in Italian valleys that nestle in the Alps.

Moving ice wears hollows in the rock below it. The hollows grow into large basins called cirques or corries. Permanently melted ice may then fill the basins. Veshensar Lake in the Himalayas was formed in this way.

The Caspian Sea also fills a giant dip in the Earth's crust. So does Lake Baykal in the USSR. This lake is so deep that it holds more water than any other freshwater lake, although there are eight lakes which have a larger surface area. Lake Baykal, with a depth of 1,940 m, is the world's deepest lake.

The life span of a lake

Many lakes last only a few thousand years. Where they have vanished, their old shores can sometimes be seen. In Scotland, near Ben Nevis, old gravel beaches high up on the sides of a mountain valley show the past levels of one vanished lake. Ice originally held back its waters. When the ice melted, the water flowed away.

In North America, a great ice sheet once trapped a huge lake now referred to as Lake Agassiz. This lake was almost the size of New Zealand. When the ice melted, most of Lake Agassiz simply ran away, leaving several smaller lakes, including Lakes Winnipeg and Manitoba in Canada.

Most lakes get filled in by rivers washing mud into them. The mud piles up until, eventually, dry land stands where the lake once lay. Lake Geneva is now one of western Europe's largest lakes. But over the next 40,000 years, the River Rhône will fill it in.

The Dead Sea, lying in Israel and Jordan, is part of the rift valley system that runs from Syria to East Africa. This deep salt lake formed in a huge dip in the crust of the Earth.

Wind at work

Many desert rocks, such as these in the Sahara, are shaped by the cutting action of wind-blown grains of sand.

Wind can move millions of tonnes of dust in a single dust storm. In dry lands, wind shifts and wears away the surface of the land. Most desert areas of the world are stony rather than sandy, because loose sand and dust have been picked up and carried far away by the action of the wind.

Bombarding rocks

You may have sat on a beach and felt the stinging force of wind-blown sand grains as they hit your skin. In deserts, wind-blown sand bombards the rocks like this. The moving sands slowly smooth and wear away the rocks, which have been weakened already by heating, cooling and sudden floods.

But sand grains are too heavy for the wind to lift very high. So wind-blown sand cuts into rock most powerfully just above the ground. Hard grains attack the lower parts of boulders until lumps of rock are left on narrow stalks, like giant mushrooms.

Cutting caves and hollows

Sand also cuts caves low down in cliffs. It attacks soft rocks more easily than hard ones. If both kinds of rock lie side by side, ridges of hard rock are found next to deep grooves, where soft rock has been worn away. Winds have scraped away so much rock in parts of the Sahara Desert, that the land there is now below the level of the sea. Egypt's Qattara Depression is the largest of such hollows. This giant 'saucer' is almost the size of Wales.

In deserts, the heat of the Sun evaporates all the water that falls as rain. In places, rain may wash together gypsum or barite minerals with sand. As the water dries up, the Sun bakes this mixture into beautiful but brittle crystals called a 'desert rose'.

Sand grains

While sand grains wear away the desert rocks, they also rub against each other. This rounds off their edges, which is why desert sands are smoother than sands formed on a beach or in a river.

Desert dust is so light that wind can pull it up into a spinning column 100 m high. These twisting pillars are known as 'dust devils'. Winds can also blow fine sand far beyond the deserts where it formed, sometimes in clouds so thick that the Sun is blotted out.

Shaping dunes

Where the wind blows strongly from one direction, sands that have been taken from one part of a desert will pile up somewhere else. These sands form shifting hills, or dunes. Dunes are created wherever the wind-blown sand is slowed down by obstacles such as stones or thorny shrubs.

Many dunes are smaller than a house. But in the Sahara Desert, where sand dunes may be hundreds of kilometres long, some are 400 m high and measure about 5km across.

Desert dust and sand is always on the move. It may be sucked up into a whirling 'dust devil', or piled into dunes. Frost, Sun, wind and sudden floods attack the rocks, carving them into many strange shapes.

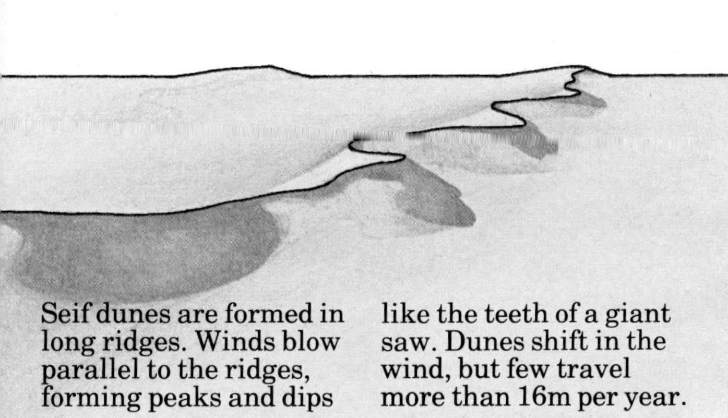

Seif dunes are formed in long ridges. Winds blow parallel to the ridges, forming peaks and dips like the teeth of a giant saw. Dunes shift in the wind, but few travel more than 16m per year.

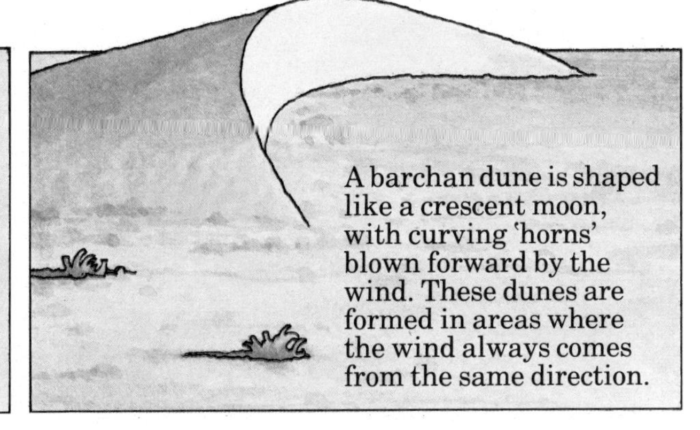

A barchan dune is shaped like a crescent moon, with curving 'horns' blown forward by the wind. These dunes are formed in areas where the wind always comes from the same direction.

Earth's skin of soil

Soil is like a skin covering the rocky surface of the Earth's crust. The skin is often very thin, but it protects the rocks below from frost and wind. It also holds nourishment for the plants on which land animals depend for their food. Without soil, all land would be an almost lifeless desert.

Thin soils of the far north

Tundra soils lie in the cold far north. Their top layer is spongy peat, made of the remains of dead moss and other lowly plants called lichens. Below the peat is bluish mud. Only a few centimetres down is a layer of permafrost – earth which is frozen solid all year round. Long plant roots cannot grow through permafrost, and melting and freezing loosens soil above, tilting many of the tiny tundra trees.

Ash-grey podzols

South of the tundra, grey forest soils cover much of northern Asia and northern North America. The grey colour of the top layer gives such soil its Russian name of podzol, meaning 'ash'. Podzols are grey because rain has washed dark, nourishing humus from the upper layer. Other substances washed down into the soil collect to form a hard layer called hardpan. Heather and needle-leaved conifers grow on podzols.

Rich black and brown soils

Black and dark brown soils are found on the open grasslands known as steppes, prairies or pampas. There is not enough rain to wash the humus deep down, so the upper layers of these soils are rich in food for plants.

Much of western Europe's farmland has brown soils. These hold the fertile remains of leaves left by the broad-leaved forests that used to cover most of the land.

Brick-red soils of the tropics

In tropical lands, heavy rain often helps to break up great thicknesses of surface rock. Here, the soil is very deep. But much of its goodness gets washed out of the topsoil. Iron deposits give tropical soils a reddish colour. Red soils support forests in wetter areas, and tropical grassland or savanna in places where the climate is drier.

Sandy soils and silt

Hot deserts have reddish, sandy soils. When rain sinks into the ground it dissolves salts in the soil. As the Sun's heat sucks up the water

The story of soil
Soil takes thousands of years to form. Firstly, rocks are weathered into smaller pieces. Lichens grow on the rock, feeding on substances that they dissolve from it.

When the lichens die, tiny organisms called bacteria break down their remains. These nourish larger plants. In time, dead plants and animals are all broken down into humus.

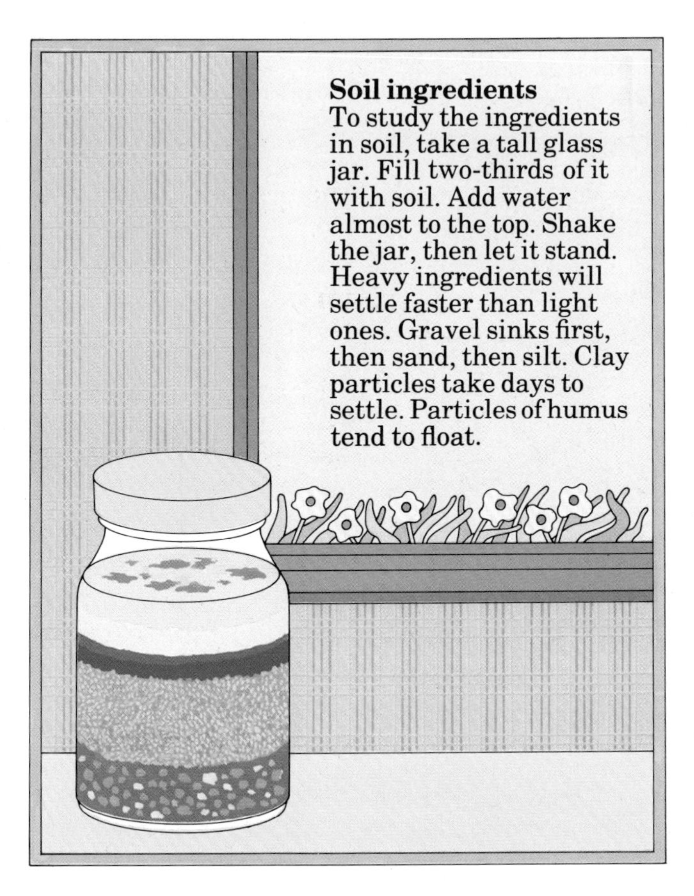

Soil ingredients
To study the ingredients in soil, take a tall glass jar. Fill two-thirds of it with soil. Add water almost to the top. Shake the jar, then let it stand. Heavy ingredients will settle faster than light ones. Gravel sinks first, then sand, then silt. Clay particles take days to settle. Particles of humus tend to float.

Many soils have several layers. In some, the top layer is rich in humus. Rain washes some humus into the next layer. The third layer is subsoil – a mixture of true soil and weathered rock broken from solid bedrock below.

again, the salts are left lying on the surface. Only a few plants grow well in these salty desert soils.

Where dust or silt from deserts and glaciers has been blown away, it may settle in layers hundreds of metres deep. Loess, the fertile yellow soil of northern China, is mostly made up of silt blown from the Gobi Desert.

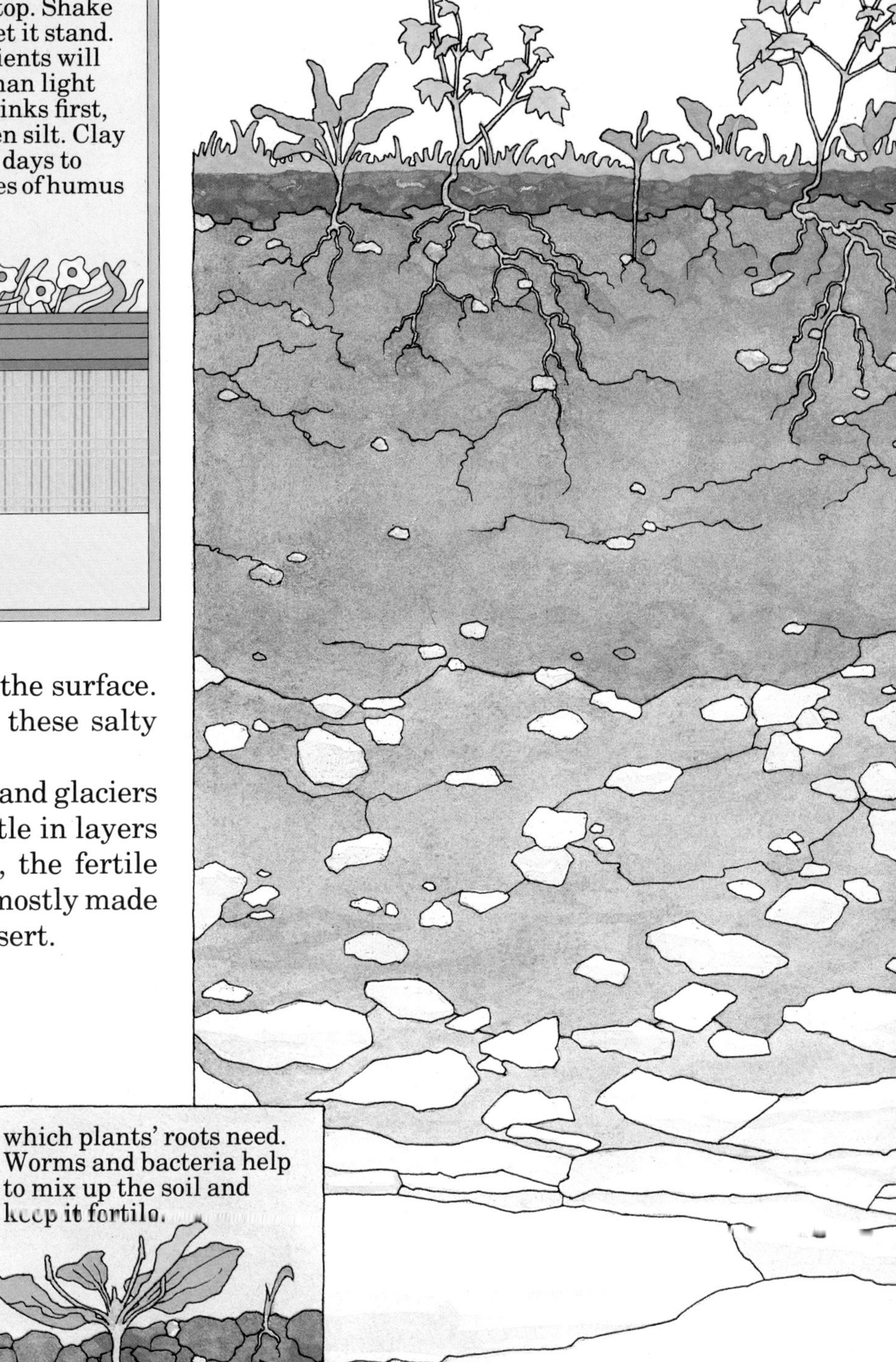

Humus, mixed with broken bits of weathered rock, forms soil. Soil also contains air and water, which plants' roots need. Worms and bacteria help to mix up the soil and keep it fertile.

Seaside battlegrounds

Coasts are battlegrounds fought over by the sea and land. The sea's attack is often sudden and exciting. Watch waves break against a shore on a stormy day, or notice how the water beats against a sea-cliff.

Destroying and building

If a cliff is made of soft, loose rocks, some often slip into the sea. If the rock is hard, waves drive air into crevices in the cliff. This happens time after time, until each crevice widens. Then chunks of rock break from the bottom of the cliff and drop into the sea. The waves pick up the fallen rocks and batter them against the cliff, so loosening other chunks of rock.

In time, the sea eats away so much of the bottom of the cliff that none remains to stop the top from falling, too. So, year by year, the cliff retreats inland.

As the sea destroys land in one place, land is growing out into the sea somewhere else. Some of the new land is built from the crushed remains of broken cliffs, which pile up as a beach of sand or pebbles.

Storms may tear away a sandy or pebbly beach. But where rivers have dropped mud in sheltered, shallow river mouths, land-building is more permanent. The roots of plants that have sprung up in the mud help to stop storms washing it away. In time, mud-flats become swamps or marshes, jutting outward from the land. Eventually, swamps may dry out and form a coastal plain.

Changing levels of land and sea

The weight of ice, or movements in the Earth's crust, can cause land to rise or fall, or the level of the sea to go up or down. Beaches, left high and dry above the ocean, show where

this has happened. Raised beaches can be seen in Scotland, Wales, New Zealand, Malta and many other places.

In the Ice Age, so much water was locked up in ice sheets that the level of the sea fell by more than 100 m. You could have walked from Ireland to Germany on dry land.

Drowned lands

But in many parts of the world, coasts drowned as the ice sheets melted and the level of the oceans rose. The North Sea filled the plain between northern Germany and Britain. Norwegian valleys, scooped out by glaciers, became steep-sided fiords.

If today's glaciers and ice sheets go on melting, many great cities will be drowned. London and Venice are already threatened, partly because both cities rest on areas of the Earth's crust that are subsiding.

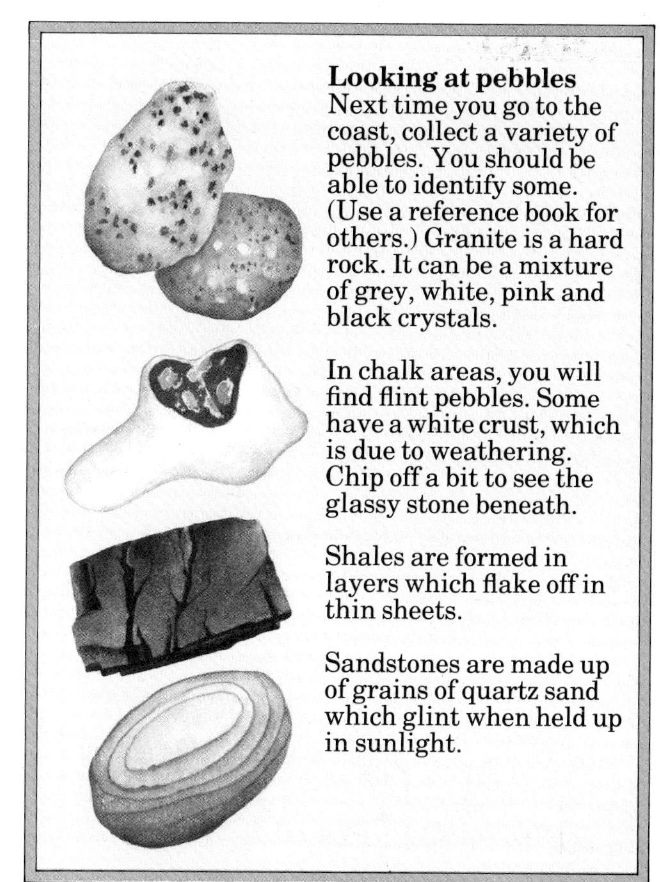

Looking at pebbles
Next time you go to the coast, collect a variety of pebbles. You should be able to identify some. (Use a reference book for others.) Granite is a hard rock. It can be a mixture of grey, white, pink and black crystals.

In chalk areas, you will find flint pebbles. Some have a white crust, which is due to weathering. Chip off a bit to see the glassy stone beneath.

Shales are formed in layers which flake off in thin sheets.

Sandstones are made up of grains of quartz sand which glint when held up in sunlight.

If cliffs of soft and hard rock lie side by side, ocean waves erode soft cliffs into bays. Hard cliffs remain poking out to sea as headlands. Mud dumped by rivers may build low coastal plains.

Oceans

If you looked down from a space craft, you would see that water covers nearly three-quarters of the Earth's surface. Most of this water lies in four great oceans. The Pacific is the largest – it could hold more than 17 countries the size of the USA. The Atlantic is the second largest and the Indian Ocean is the third. The Arctic Ocean is the smallest of the four.

The movement of water

Ocean water is always moving. Currents of wind-blown water flow like rivers across the surface of the sea. Below them, other currents often flow in opposite directions. Winds stir the surface into waves. The Moon's gravitational pull upon the spinning Earth keeps huge, slow-moving waves called tides travelling round the oceans.

The effect of temperature

Heat and cold also keep ocean water on the move. The Sun's heat draws water up from the ocean surface into the air, as water vapour. When water vapour cools it turns to rain that falls back into the sea, or runs into rivers that flow into the sea. Where sea-water freezes, floes (or ice islands) are formed. These drift far across the cold oceans that surround the north and south poles.

Oceans of salt

Oceans are not only made of water. For thousands of millions of years, other substances have been collecting in the sea. Some chemicals escape into the water from undersea volcanoes, and salts are washed off the land by rivers. When the Sun's heat draws sea-water into the air, salts are left behind. In these ways, salts have been collecting since the oceans began. This is why sea-water tastes so salty.

The changing oceans

Like the continents, the oceans have changed their shapes and sizes. When the huge mass of land called Pangaea broke up, the Arctic, Atlantic and Indian Oceans were born in the gaps created between continents as they drifted apart.

Where two continents came together, they closed, or partly closed, the sea that used to separate them. This happened when India collided with Tibet. Before that, the Mediterranean was part of a much larger sea that stretched as far as China.

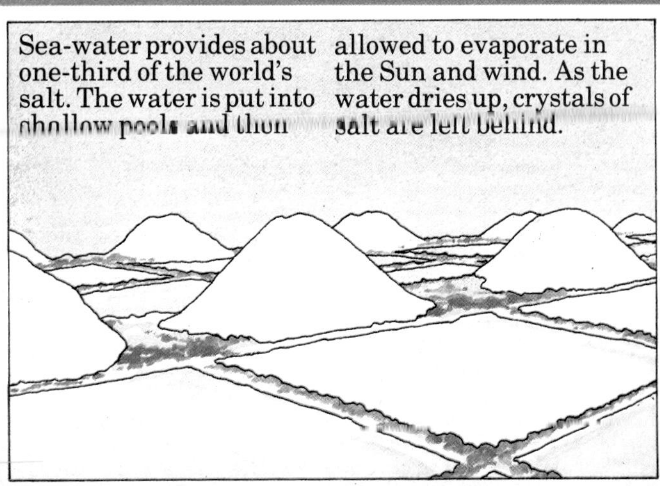

Sea-water provides about one-third of the world's salt. The water is put into shallow pools and then allowed to evaporate in the Sun and wind. As the water dries up, crystals of salt are left behind.

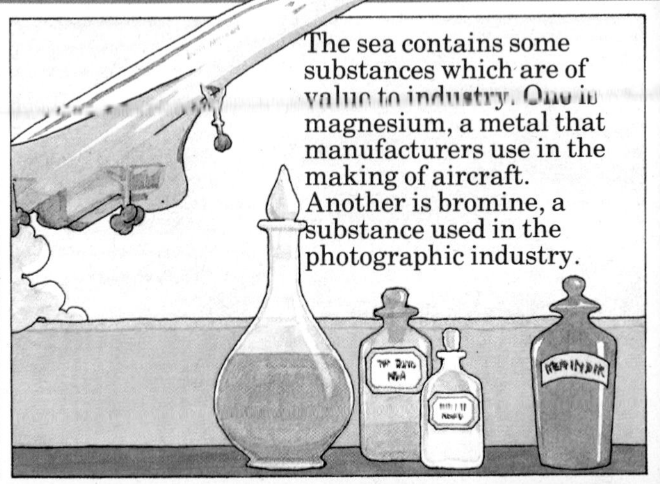

The sea contains some substances which are of value to industry. One is magnesium, a metal that manufacturers use in the making of aircraft. Another is bromine, a substance used in the photographic industry.

Pressure at depth
Expert pearl divers can descend to 30m, where water pressure is four times greater than normal air pressure at sea level.

Water pressure increases with depth. At 1,133 m, it is more than 120 times that of sea-level air pressure. A sperm whale can dive to this depth.

Humans need very special equipment to go so deep. In 1960, two men in a bathyscaphe dived to the amazing depth of 10,900 m (nearly 11km).

Earth is a very watery planet, as a view of it from above Tahiti shows. Almost all that can be seen is the Pacific Ocean, the largest and deepest of all oceans.

The ocean floor

Like dry land, the ocean floor has mountains, plains and valleys. You would see all these if you could drain the Atlantic Ocean, and walk across it from Europe to America.

The continental shelf

At the beginning of your journey, you would come to a platform called the continental shelf. Such platforms lie beneath the shallow waters that surround most continents. Parts of these shelves were once dry land. They can be recognised by their drowned beaches, cliffs and river beds.

Plunging downward

The continental shelf ends with a long, steep slope which plunges downward to about 3,800 m. In some places, water flowing from rivers has carved deep valleys in this huge wall. Lower down, the slope becomes less steep. By now, you would be squelching across soft ooze. This muddy carpet is mostly made up of the shells of tiny sea animals. They once lived in the surface waters, but sank to the sea-bed when they died.

Next, you would arrive at an extremely deep underwater plain. Much of this abyssal plain has a thin coating of red clay. Its tiny particles were largely blown or washed into the sea from the land, thrown out by volcanoes and dropped by melting icebergs.

Underwater mountains

In the middle of the Atlantic, you would cross the Mid-Atlantic Ridge. This great chain of underwater mountains is built of lava, which poured from the crack that opened between two slowly separating crustal plates. Fresh lava still comes up and sticks to the edges of the plates as they move apart.

A vast chain of ridges, or underwater mountains, circles the Earth. Where peaks break the surface they form islands.

Trenches are deep gashes in the ocean floor. They occur where two plates collide, and one is dragged into the mantle.

The continental shelf, parts of which were once dry ground, is covered by sands and muds washed off the land by rivers.

This creates new ocean floor. Because the ocean floor is spreading, volcanic mountains born on the crest of the Mid-Atlantic Ridge move gradually outward from the middle of the ocean. This partly explains why some islands now stand on one side of the ocean instead of in the middle, where they started.

Trenches

Just before reaching the West Indies, you would cross a deep, narrow trench. Spreading ocean floor arriving here gets dragged down into the Earth's mantle. Sea-water that goes with it later escapes from volcanoes as steam. Volcanoes also give off water that has been locked up in rocks below the crust.

No ocean floor lasts longer than 200 million years. So the oldest ocean floor is almost 20 times younger than the oldest land rocks, which go back 3,800 million years.

Abyssal plain clay is a mixture of dust, sharks' teeth and whales' ear bones. Great pressure dissolves everything else as it falls.

A rain of dead and dying sea plants and animals is always falling to the ocean floor. Most are dissolved on the way down by the tremendous water pressure. So it may take 20,000 years to build up a layer of ooze just 2.5cm thick.

Layered rocks

Ocean floors are like giant dustbins, as they collect the remains of dead sea plants and animals, and particles of broken rock washed or blown off the land into the sea. Such things which settle on the ocean floor are called sediments. Sediments can also form in lakes and deserts and under glaciers.

How sedimentary rocks form

Sedimentary rocks are made from layers of sediment that hardened over thousands of millions of years. Huge areas of the world are covered by sedimentary rocks, and some layers are immensely thick. South-east England has a layer of chalk 300 m thick.

Many sedimentary rocks are built up from the broken remains of igneous, or 'fiery', rocks. Sandstones, for instance, are mainly made of grains of quartz which come from weathered granite. Clays and mudstones are made of tiny particles of quartz and mica. Grit, conglomerate and breccia are three natural kinds of concrete. Each is formed from sand grains and pebbles that have been 'glued' together by certain chemicals.

Limestones are sedimentary rocks usually made from the hard parts of millions of plants and animals that once lived in ancient seas. Some limestones come from corals. Others hold countless sea shells. Fossils are found preserved only in sedimentary rocks.

Natural production of fuels

Coal, oil and gas are three of our most useful fuels. All of them are produced during the formation of sedimentary rocks.

Coal is a sedimentary rock made of carbon. The carbon in coal is the remains of plants that formed low, swampy forests hundreds of millions of years ago. The coal-forest plants grew as tall as trees. Yet many were related to today's club mosses – tiny plants that you walk on without even noticing.

As the coal-forest plants died, they fell into their swamps. In time, the swamps grew deep

Fossils
Fossils are the remains of prehistoric plants and animals preserved in rock. These include the shells or skeletons of animals, and the woody parts of plants. Firstly, a dead animal or plant sinks to the sea-bed.

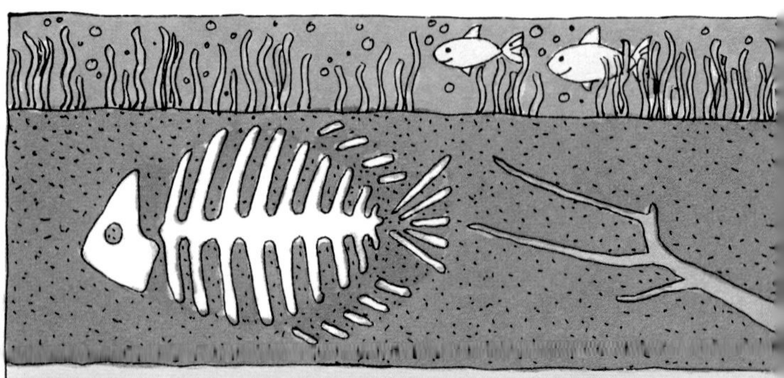

In animals and plants, the soft parts soon rot away, leaving a skeleton. This gets covered by particles of mud or sand. Then the animal or plant skeleton slowly begins to disintegrate.

This rock formed in the sea about 160 million years ago. Ammonites once lived in the coiled, fossilized shells.

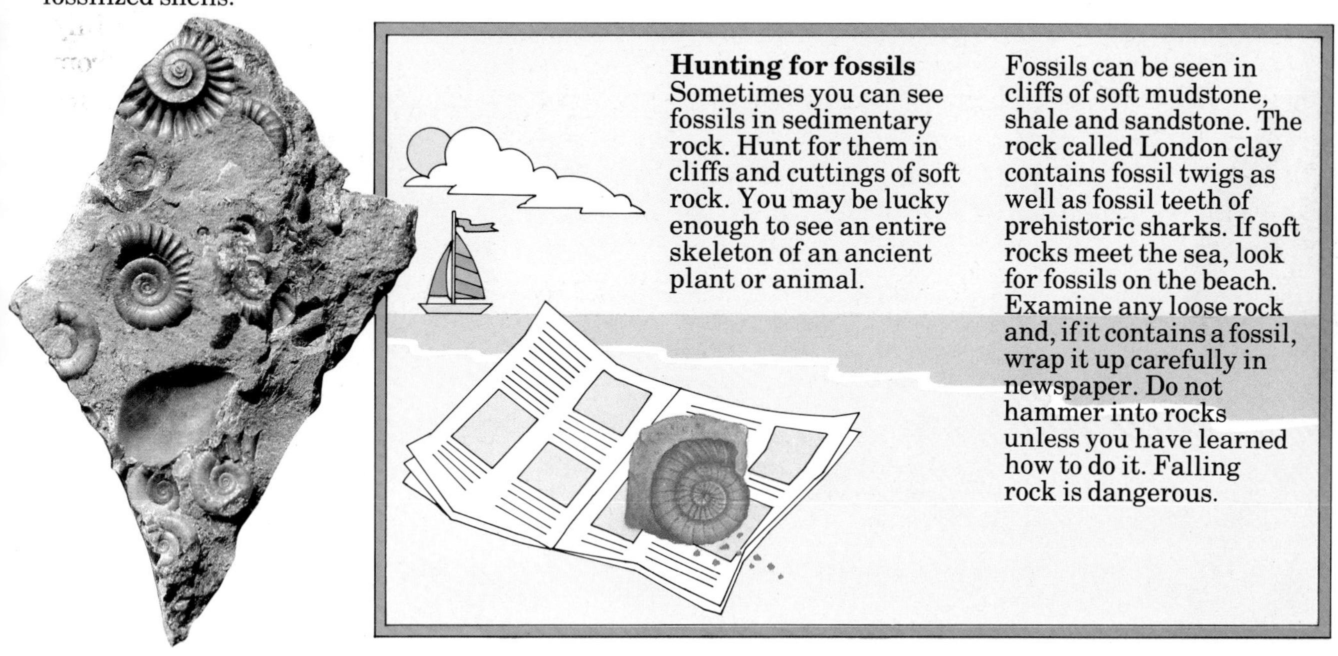

Hunting for fossils
Sometimes you can see fossils in sedimentary rock. Hunt for them in cliffs and cuttings of soft rock. You may be lucky enough to see an entire skeleton of an ancient plant or animal.

Fossils can be seen in cliffs of soft mudstone, shale and sandstone. The rock called London clay contains fossil twigs as well as fossil teeth of prehistoric sharks. If soft rocks meet the sea, look for fossils on the beach. Examine any loose rock and, if it contains a fossil, wrap it up carefully in newspaper. Do not hammer into rocks unless you have learned how to do it. Falling rock is dangerous.

in dead plants. Then the sea invaded the swamps and covered the plants with sand or mud. Later, the sea retreated and a new forest grew upon the mud. This happened many times, until there were several layers of dead plants. The weight of the sand and mud above them squashed these layers and so helped to compress them into coal.

Oil forms when millions of dead sea plants and animals sink to the bottom of the ocean and turn to slime that becomes buried by layers of mud and sand. After millions of years, the slime, mud and sand are squashed. The lower layers of mud and sand turn to shale and sandstone rocks, while the squashed slime turns to oil and gas. Oil and gas fill the tiny pores, or gaps, in porous rocks such as sandstone. Inside these rocks, the oil floats on top of salt water, and any gas floats on the oil, as gas is lighter than oil or water.

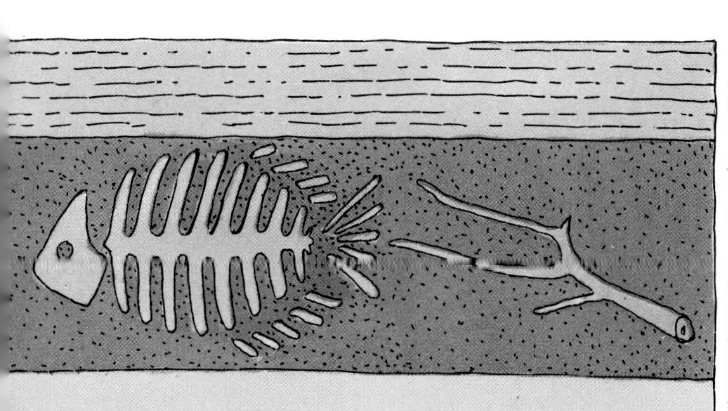

The gaps left are filled by minerals brought by water seeping down from above. So stone replaces bone or wood. The layers of mud or sand lying above the fossil harden into rock.

In time, underground movements raise the rock above the sea. Wind and rain wear it away until the fossil shows up on its surface. Millions of years may pass between an animal or plant dying, and its fossil being discovered.

The future

People have not lived on Earth long enough to have seen continents collide or split apart, mountains grow or rivers wear them down. Yet great natural changes have happened to the Earth since Man's ape-like ancestors first appeared, 6 million years ago. At one time, for instance, the Mediterranean Sea dried up and all of the three northern continents were linked by land.

Changes made by people

In modern times, humans have left their own mark on the Earth's crust. By blocking rivers, engineers have made huge lakes. (Lake Volta in Ghana is more than three times larger than Luxembourg.) They have changed the direction in which some rivers flow, joined oceans by canals, and tunnelled through mountains and beneath the sea.

Quarrymen have dug enormous pits and built hills with the rock waste. Dutch engineers have won land from the sea by raising walls across bays and pumping out the water. On the darker side, careless farming has increased soil erosion and turned some fertile land into stony desert.

Plans on a vast scale

Great changes may still lie ahead. One idea is to close off the Mediterranean Sea by building dams. A shrinking Mediterranean would not only change the landscape, but possibly the climate as well.

A more realistic scheme is the Russian one to reverse the Ob and Yenisei Rivers. Canals, cut from the lakes where the rivers rise, would carry their waters to desert areas. Even the Caspian Sea, which is drying up, could be fed by these two rivers.

If the Mediterranean Sea were dammed and began to dry up, land would join Sicily to Italy. But such major engineering could cause volcanic activity and earthquakes, by reducing the weight of water on the sea-bed.

The physical Earth

Statistics of the Earth

Circumference at the equator	40,075 km
Thickness of the crust	6km under sea; 40km under land
Thickness of the mantle	2,900 km
Thickness of the outer core	2,240 km
Thickness of the inner core	1,200 km
Estimated temperature at the centre	2,800°C
Proportion of land to water	30% land; 70% water
Number of active land volcanoes	455
Estimated number of active underwater volcanoes	80

Earth's physical records

Largest continent	Asia	54,527,000 sq km
Smallest continent	Australia	7,614,500 sq km
Largest island	Greenland	2,175,000 sq km
Largest ocean	Pacific Ocean	165,250,000 sq km
Smallest ocean	Arctic Ocean	8,766,560 sq km
Deepest part of ocean	Marianas Trench, Pacific Ocean	10,900 m
Largest salt lake	Caspian Sea, USSR and Iran	371,800 sq km
Largest freshwater lake	Lake Superior, Canada and USA	82,350 sq km
Deepest lake	Lake Baykal, USSR	1,940 m
Highest mountain	Mt Everest, Himalayas	8,848 m
Largest desert	Sahara Desert, Africa	8,400,000 sq km
Longest river	Nile River, Africa	6,670 km
Longest glacier	Lambert Glacier, Antarctica	514 km

Some common rocks

Igneous
granite
diorite
gabbro
obsidian
andesite
basalt
Sedimentary
limestone
sandstone
clay
shale
grit
conglomerate
Metamorphic
marble (from limestone)
schist (from shale)
gneiss (from granite)
slate (from shale)
quartzite (from sandstone)
greenstone (from basalt)

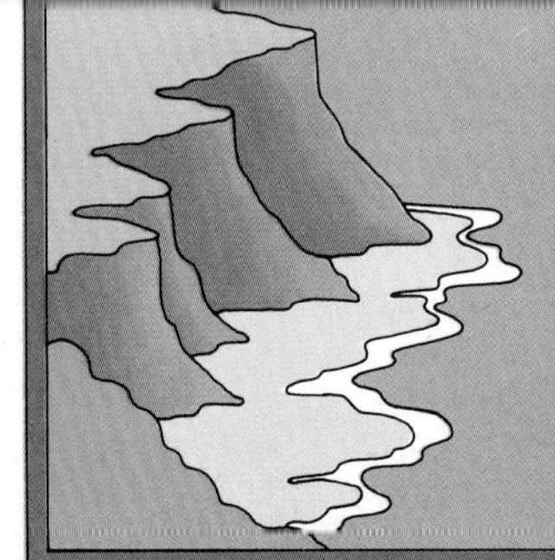

Mini museum
You can learn quite a lot about the history of the land near where you live by making drawings and notes of the countryside. Notice where rivers have shaped valleys or where the sea has worn away cliffs. Examine the rocks to see whether they are soft or hard, and if they contain crystals. Take home some samples of the different rocks and pebbles you find. Label all the drawings and objects and arrange your collection on a table or window sill.

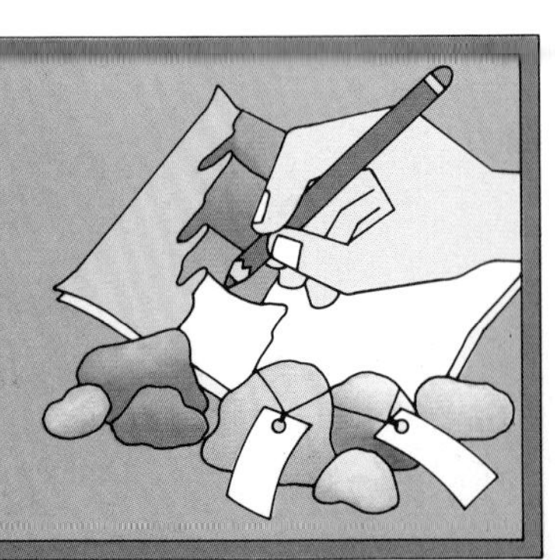

Index

Page numbers in *italic* refer to illustrations.

Earth's time scale
200 million years ago to 40 million years in the future

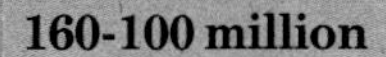

200-160 million

Pangaea begins to break up. 200

The first mammals appear. 190

Africa and North America separate and the modern oceans begin to appear. 180

India splits from Australia and Antarctica. 170

160-100 million

The first birds appear. 150

Africa and South America begin to separate. 135

North-western Europe is covered by a shallow, chalky sea surrounded by deserts. 120

100-25 million

Europe and North America are moving apart. 70

The Rocky Mountains and the Andes are rising. 65

The dinosaurs die out. 65

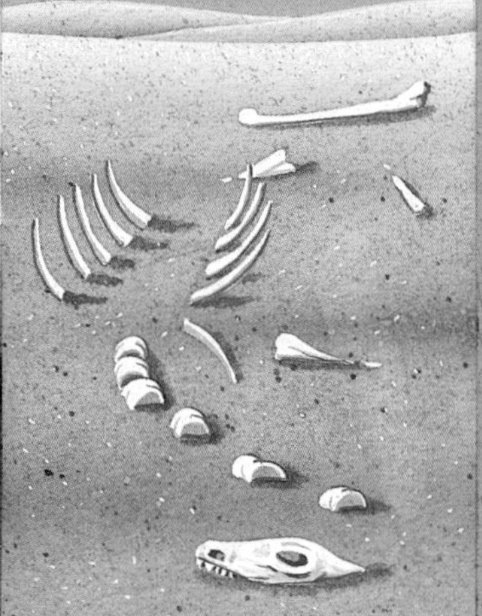

Sea separates North and South America. 60

The European Alps are pushed up when Italy glides slowly into southern Europe. 40

25-5 million

The first apes appear. 25

The Mediterranean Sea repeatedly dries up and fills again. 20-5

Central America begins to rise from the sea. 12

Land links North and South America. 7